THE
MADAME
CURIE
COMPLEX

THE WOMEN WRITING SCIENCE PROJECT

In collaboration with the National Science Foundation, the Feminist Press announces Women Writing Science, a series of books that bring to life the story of women in science, engineering, technology, and mathematics. From biographies and novels to research-based nonfiction, Women Writing Science will celebrate the achievements of women scientists, explore their conflicts, address gender bias barriers, encourage women and men of all ages to become more scientifically literate, and embolden young women to choose careers in science.

Other books in the series

WOMEN IN SCIENCE: THEN AND NOW
Vivian Gornick

BASE TEN
Maryann Lesert

THE MADAME CURIE COMPLEX

The Hidden History of Women in Science

JULIE DES JARDINS

THE FEMINIST PRESS
AT THE CITY UNIVERSITY OF NEW YORK
NEW YORK CITY

Published in 2010 by the Feminist Press
at the City University of New York
The Graduate Center
365 Fifth Avenue
New York, NY 10016
Feministpress.org

First printing March 2010

Cover and text design by Drew Stevens

Publication of this volume is made possible, in part, by funds from the National Science
Foundation and Sara Lee Schupf.

Library of Congress Cataloging-in-Publication Data
Des Jardins, Julie.
The Madame Curie complex : the hidden history of women in science / Julie Des Jardins.
    p.   cm. — (Women writing science)
    ISBN 978-1-55861-613-4
1. Women scientists—Biography. 2. Women engineers—Biography. 3. Women in
science. 4. Women in engineering. 5. Sex role—United States—History. 6. Spouses.
I. Title.
    Q141.D44 2010
500.82—dc22

                                                                    2009037794

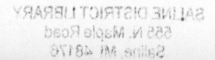

To my dad,

my Bella,

and a meeting of the minds

Contents

Introduction: Through the Lives of Women Scientists

SOME FATHERS TELL THEIR KIDS ABOUT MYTHICAL HOME RUN HIT-ters who won the World Series, or of courageous expeditions to Antarctica or the moon. My father's story to me was about Enrico Fermi and the scientists who created a nuclear chain reaction beneath the stands of the football stadium at the University of Chicago during World War II. As a lifelong Chicagoan, Dad preferred this story to those about heroic men at Los Alamos. It was at this moment under Stagg Field, he believed, that brilliant men had convened for some greater scientific purpose. The only story as awe inspiring was the one he told of the Founding Fathers, who converged on Philadelphia in 1776 to establish, as Dad put it, the most enlightened democracy in the world. Both stories were so often told in our house that I could even remember the pauses built into their telling for dramatic effect. Whether he knew it or not, the message Dad conveyed was not only that science, like statescraft, bred greatness, but also that those talented enough to change the course of history were inevitably male.

My father was in awe of the great men of politics and science, and I wanted to awe my father. I never aspired to run for office, but I did have fleeting thoughts about becoming a scientist. I started out on the right path, and early on I proved to be a whiz at math. In elementary school I performed arithmetic leaps and bounds better than my peers. I remember my fourth-grade teacher walking me over to the accelerated fifth-grade class while the other kids grappled with concepts I had already mastered. My excellence in math was short-lived, however. In sixth grade I was ahead by two grade levels, and by seventh and eighth grades ahead by only one. In high school I hit my wall, turning in a mediocre performance in advanced algebra and then doing miserably in trigonometry. Invariably, boys were involved, for the ones on which I had designs didn't turn up in my history classes, but in math classes, where I deferred to their seeming expertise. In my senior year I opted out of calculus altogether. It infuriates me that I had succumbed to math anxiety at precisely the predictable moment for adolescent girls. Then, although I performed brilliantly in advanced biology, I had so little confidence in math that I avoided other science classes. In the one chemistry course I was forced to take, the teacher seemed openly hostile to girls, so I didn't approach him for help. Although I could have pursued biology in college, I decided that I didn't have a mind capable of serious science.

My dad once confided to me that he thought things would have been different had I been a boy; he sensed that I had fallen prey to the prevailing tides. In his honor, here are the stories of scientists, ones that I hope will also inspire awe. Of course, they are necessarily different from his, since I am asking new questions about heroism. I am asking about the girls who did what I didn't, who felt social pressures not to pursue science but did anyway. I am asking about the lives they led once they became women and certifiable scientists, and I am asking about how it was that many of the best female scientific minds were ignored largely because they never were certified as "scientists." These stories describe scientists my father never heard of, though some of them might have been under Stagg Field had the culture of science been different.

This book is not overly laudatory of female accomplishment in science, but it isn't a victimology either. To pity women of the sexist past or to celebrate women's progress in the enlightened present is to write without context. To find compelling stories in the twentieth century, I had to leave Stagg Field and sometimes professional labs and institutional science altogether. I had to write outside the narrative conventions of "great" history, too, for when you boil them down, they are masculine to the hilt and perpetuate the invisibility of women I aim to remedy. Women can never be the true heroines of "great science," just wannabes and impostors, since the cast of such science is male by default. Twentieth-century science has been buoyed by myths that create prestige in masculine terms, and thus it is not enough to provide a laundry list of the women who nevertheless managed to earn entrance and accolades into its elite circles. Such stories could never be anything more than fragmentary and compensatory history, filling yet another volume of "exceptional women who performed outside their skins." I've had to question the concept of scientific heroism and define it by other means. The drama of women's stories lies less often in extraordinary discovery than in overlooked details—details that pertain to women's scientific work but also to their private lives, since both are inextricably linked. Feminists have long insisted that the personal is political, and I extend that truism to assert that the personal and professional, private and scientific, are intertwined in these pages.

To tell a connected story of women in science, my historian's lens pans out to reveal the contours of women's science careers generally, but it zooms in to look at representative figures up close—some iconic, and others not successful in the traditional sense. Marie Curie, Jane Goodall, Rosalind Franklin, Evelyn Fox Keller, Lillian Gilbreth, Cecilia Payne-Gaposchkin, Dian Fossey, Annie Jump Cannon, Biruté Galdikas, various Nobel laureates, and women of the Manhattan Project appear separately and as part of a whole to shed light on the ways American science has been experienced and imagined over the past century. While it might sound strange to bring together scientists from the lab and the field, the Nobel podium and the back room, the physical and life sciences, academic and popular venues from the turn of the century and today, I tell their individual and collective stories to show how cultural ideas of gender have shaped the methods, structure, and meaning of science itself.

Some may wonder how intrusive gender ideology has been of late, for sheer numbers seem to indicate that women have made progress in scientific institutions over the past century and that the trend continues. As I scan the admissions website of the Massachusetts Institute of Technology (MIT), I see that it currently boasts an undergraduate population that is nearly half female, and even a female president. When seen in the light of the ratios of women to men in the U.S. Naval Academy's class of 2009 (19 to 81 percent), it seems that the culture of MIT and science generally has already changed in ways that the American military—yet another historically male bastion of prestige— has not. And yet I couldn't help but also notice the data that was conspicuously missing on the MIT website. I wonder about the fields in which female students succeed most, and what their outcomes are in graduate school or the job market after college. The website boasts of 197 female faculty members, but fails to inform us how this compares with numbers of men or at which levels of promotion women congregate. Although it pays homage to MIT's first female doctorate holder in chemistry, the site fails to convey that she never achieved the pay or promotion of her male peers and to mention that she was relegated to the "women's work" of domestic science. More important, it indicates nothing about the types of scientific questions pursued at MIT. How have women, whatever their numbers, actually changed the culture of the institution and the social ramifications of the work done there?[1]

I'm guessing that the data on female bodies walking the grounds at MIT looks better to prospective applicants than some of the data I want to know, but history tells me that deeper meaning lurks in these and other cultural silences beyond *scientific* institutions. In recent decades we've applauded the heightened female presence in American politics, for instance, and yet antifeminist legislation gets passed under women's watch all the time. The 2008 vice presidential run of Sarah Palin underscores the point that peopling institutions with women will be no hard-and-fast way to change the sexist culture that pervades them. And thus, while no one disputes that more women occupy institutional science than ever before, I think that we need to pay more attention to the *gendering* of scientific culture in the end. Luckily for us, the personal and professional lives of women provide a lens for seeing this process over time. The stories in this book will enlighten, and at the same time they may also inspire.

I have written these stories as the problem of "women in science" continues to rear its head in twenty-first-century popular discourse. Naively, I had at first assumed that "the problem" was no longer a problem at all and that a general consensus had been reached: that women are adept at science and that they simply need the opportunity to prove it. I've since learned, however, that present attitudes are more complex. Most people I know think that women can be good scientists; but they also believe, consciously or not, that what makes women good scientists is the extent to which they deny their true selves to think like men. Knowing this, I'm afraid the "women in science" problem is far from licked. In January 2005 Larry Summers, then the president of Harvard University, managed to stir the pot when he remarked to an audience including scientists that women's relatively poor showing in science could be linked to innate biological difference. When I heard the early media reports I was outraged, but transcripts reveal more ambiguity in Summers's statements than what the media first claimed. He attributed women's lower numbers in science to any number of factors, discrimination included. The essence of what he said is probably less relevant than the charged reaction that ensued, since it tells us so much more. Women's position in science will never be resolved so long as gender itself remains a hotly contested social problem.[2]

It seems that everyone has opinions on whether or not women have scientific ability, and some opinions are more informed than others. My intention is simply to bring the past to bear on them. Few of the women scientists who feature in this book entered these debates directly, but their work speaks for itself, dispelling claims that men are inherently better than women at science. With their limited access to the professional training, resources, mentoring, and social networks of men, women have occasionally asked different questions, used different methods, and come up with different explanations for phenomena in the natural world. For many, their marginal status in institutional science afforded their altogether different relationship to the nature they observed. As some became stewards and advocates for women, animals, and the environment, they expanded definitions of appropriate scientific work in the twenty-first century.

Of course the title of this book begs the question of Marie Curie's effect on these stories about American women and science. She wasn't American, yet the sheer number of American books and films written about her suggests that she is a presence we cannot ignore. In my adolescence I remember reading one of these biographies and thinking that if she could immigrate to another country and stave off death to study science, I could stick out Intro to Physics. Many biographers of women scientists treat her as the first and most inspiring female role model of the modern era, and I, too, introduce her first in the stories that follow, but not for the same purpose. No doubt, her work helps to make the case that science is *made* to be manly rather than being inherently so. But Curie's myth haunts these pages and the psyches of succeeding generations of women more completely than her real-life example, for it has both empowered and stigmatized women, liberated and constrained them, often at the same time. The historian Margaret Rossiter noted an inferiority complex in women after Curie's tours of the United States in the 1920s, and for generations the Curie complex has continued to allow men to disqualify women—and women to disqualify themselves—from science. Women scientists have felt as though they cannot measure up to Curie, and of course how could they, when this mythical measure of female competence has morphed in the American mind over and over again? It's like trying to hit a moving target.[3]

I have organized my stories of women around three historical "hubs" that coincide roughly with the first, middle, and final third of the twentieth century. The first is characterized by scientific professionalization, the second by World War II and the postwar veneration of masculine science, and the third by the rise of second-wave feminism. Regardless of the historical backdrop, the stories contained herein speak to questions about women in science that continue to be relevant in the twenty-first century: Can women be both "womanly" and "scientific?" How do women balance their home life and scientific careers? How do men and women collaborate in the lab? Are certain scientific environments better for women than others? Do women "do" science differently from men? Is there such a thing as "feminist" science? Rarely did the women in these pages ask such questions explicitly, and yet their stories make us reflective about possible responses.

In Part I, "Assistants, Housekeepers, and Interchangeable Parts," the masculinizing effects of professionalization redefine women's relationship to science in ways that reverberate today. The very definition of "professional science" relies on its antithetical concept—amateur science, defined in its methods and orientation as quintessentially domestic. In professional settings women performed science in relative obscurity or occupied niches in pedagogy, popularization, and domestic research. When they made inroads into men's labs it was almost always as assistants, technicians, and helpmeets. In Chapter 1, Marie Curie found herself negotiating American expectations of womanhood when she toured the United States to obtain radium for her research in the 1920s. Her admirers could not idolize her as a scientist, except to assume that domesticity and maternal benevolence motivated her work. In the American press she did not discover radium for the sake of science, but rather to minister to victims of cancer, as good women would.

In Chapter 2, Missy Meloney, the American journalist who organized Curie's radium tours, asked the Nobel Prize winner to nominate the industrial engineer Lillian Gilbreth for a Nobel Prize in 1932.[4] Curie told Meloney that she would oblige if she knew which Nobel Prize to nominate her for, for Gilbreth was not a chemist, physicist, or medical scientist. Indeed throughout her life Gilbreth defied the facile categories of others. She was mother and wife—and hence a domestic woman, and yet she also made a living in the most virile of scientific fields. Male industrial engineers sometimes ignored her, if not her work, and yet she became one of the most popularly recognized American women of the twentieth century. What makes her story emblematic of so many other women scientists is that few of her admirers recognized her as a scientist at all. Women scientists in these years often went by other names, whether they liked it or not. In Chapter 3, for example, the first generations of women "computers" at the Harvard Astronomical Observatory suffered similar sexist semantics: though they worked in university buildings where serious research presumably took place, they were tolerated as "scientific housekeepers"—custodians of records, cleaners of equipment, boosters of morale for the important men in their presence. Records viewed today amply demonstrate that the work of the Harvard computers was often more interpretive and groundbreaking than its categorization as "busy" or "domestic" suggested. Still, Annie Cannon's embracing of her status as a womanly helpmeet to male astronomers won accolades for her as well as the ability to establish funding for women in future pursuits of science. When we examine the career path of her younger co-worker Cecilia Payne-Gaposchkin, we see advantages and drawbacks to both assimilation in masculine professional culture and capitulation to traditionally female positions in science. In this instance history complicates as much as it clarifies.

Part II, "The Cult of Masculinity in the Age of Heroic Science"—addressing the period of the middle third of the century—describes a time when science enjoyed its highest esteem in American life as well as its most virile identity. While Albert Einstein was a popularly recognized figure in the 1920s and 1930s, during and after World War II, when the ramifications of atomic science became widely apparent, the image of the male physicist became still more venerated. Robert Oppenheimer and Enrico Fermi became larger than life in the American mind, revered in ways no woman scientist has been in the postwar era. In Chapter 4, cultural amnesia regarding women of the Manhattan Project is seen as one consequence of the masculine mythology of these years. Another is the devaluation of women's Nobel-caliber science. Chapter 5 illuminates how Gerty Cori and Maria Goeppert Mayer benefited from their close proximity to scientific men and were acknowledged as Nobelworthy scientists, and yet being married to prominent men also meant tolerating decades of underemployment. They treaded close enough to men to be supported and protected, while keeping an arm's distance so as also to be distinguishable in their own right. The Nobel's snubs of single women Lise Meitner and Rosalind Franklin represent more cautionary tales, but the qualified success of Nobel winner Rosalyn Yalow, treated in Chapter 6, suggests that succeeding in institutional science in the postwar years required a crafty adaptation of the masculine role for legitimacy and the womanly role for acceptance.

Part III, "American Women and Science in Transition," covers scientists during the last third of the twentieth century, with the rise of second-wave feminism. In this period, professional science began to shed its masculine trappings, at least in the work of Louis Leakey's "Lady Trimates": Jane Goodall, Dian Fossey, and Biruté Galdikas. In Chapters 6 and 7, I relate how Barbara McClintock, Rachel Carson, and others seized on and facilitated this moment of flux. They rejected man's presumed dominance over nature as well as beliefs in "masculine" objectivity to embrace scientific work as popularizers, naturalists, and advocates—bearing willingly, even boldly, labels that had long been placed on women so as to reduce them to scientific "amateurs."

In this postatomic age, both male and female scientists have come to understand the importance of bringing nonscientists into their work, for decisions about nuclear bombs and gene replication have social ramifications that cannot remain secret. Women's ability to make science accessible takes on new significance going into the twenty-first century.

I concede that my pigeonholing women scientists into tidy historical categories—of which they themselves were never aware—may be fraught with problems. But I trust that this volume makes clear that I want, not to reduce women, but rather to position them to be seen anew, perhaps even for the first time. I also want to create a useful framework through which to see many facets of gender in science. Several women practiced science across my sometimes arbitrary chronological divisions, and certainly women's lives and thoughts transgress the generalizations I have made about them as members of generational, disciplinary, or ideological cohorts. Still, there is a larger picture that the individual stories of women can clarify when the narratives are put together and seen as a whole. Suddenly the historical lens changes too, and we can see women practicing science when it wasn't in the lab, the university, or under Stagg Field. We can even flesh out the one woman who did stand witness to the nuclear reaction my father talked so much about. Her name was Leona Woods Marshall, later Leona Marshall Libby. In this volume, she is more than a footnote: she is a fresh pair of eyes on the culture of American science.

Notes

1. Http://www.mitadmissions.org/topics/life/women_at_mit/; http://www.usna.edu/Admissions/classprofile.htm, April 6, 2009.

2. "The AAUP's Committee on Women Responds to Lawrence Summers," *Academe*, July/August 2005, 59; "Summerstime, and the Living Ain't Easy," *Economist*, February 26, 2005 (http://www.the-economist.com/research/articlesbysubject/displaystory.cfm?subjectid=2743324&story-id=E1.PGJVJPP); Amanda Ripley, "Larry Summers," *Time*, April 18, 2005, 104.

3. Margaret Rossiter, *Women Scientists in America: Struggles and Strategies to 1940* (Baltimore: Johns Hopkins Press, 1982), 126–27, 130, 216.

4. Marie Curie to Missy Meloney, January 29, 1931, Box 1, William B. Meloney-Marie Curie Special Manuscript Collection, Columbia University Libraries, New York, NY.

I Assistants, Housekeepers, and Interchangeable Parts: Women Scientists and Professionalization, 1880–1940

THE YEAR WAS 1921. THE PHYSICIST ROBERT MILLIKAN ANNOUNCED that the nation's expanding technostructure required better "selection and development of men of outstanding ability in science." His call for "men" was not accidental; he imagined the best candidates to be masculine, rugged types and likened them to "explorers" in search of "nature's gold." Thomas Alva Edison, holder of more than one thousand American patents and the most widely recognized scientist in the country, embodied this enterprising pioneer as he attempted to recruit young men in his mold to develop technologies in his research facility in West Orange, New Jersey. Edison wouldn't settle for small-time thinkers; his team of "A-class men" would be drawn from applicants who passed a rigorous test he had devised. But his recruitment efforts proved disappointing: of the five hundred men who applied for positions, only 6 percent passed his exam. His own son failed to earn a place in his ranks of A-Class men.[1]

Test takers grew defensive; they were not common street folk, but graduates of science programs in the most prestigious universities in the country. The problem lay not with them, they claimed, but rather with Edison's arbitrary criteria for gauging scientific competence in the modern age. Some of the questions on his exam were what they were used to: "What pinch pressure at the driving wheels does a 23-ton locomotive require when drawing a load of 100 tons on level track?" Others, however, seemed highly irregular for testing scientific competence: "Who was Leonides?" "What is the name of a famous violin maker?" "What is felt?" One stumped job applicant wondered, "How many $10,000 per annum men . . . could have answered 50 percent of those tomfooleryisms." Another dismissed the test as "vulgar," an insult to his educated sensibility. "Who cares who wrote 'Home Sweet Home,'" a college graduate lashed out. "We are in an age of specialization, and men are being trained to do things in certain lines of work that do not allow them to waste time and gray matter on general knowledge that can be had by referring to an encyclopedia."[2]

Not all reactions to Edison's questions were defensive; some thought that the test proved just how "amazingly ignorant" college men had become. "I think that any man who cannot give a prompt answer to 75 percent of the questions at least is lacking in education, and, if a college man, had wasted his time in college," asserted an anonymous reader of the *New York Times*. Another reader thought the questions answerable "by any well-read and average intelligent man or woman," regardless of college credentials. Some thought it refreshing that Edison looked for men who didn't have "one-track minds," who sought to expand their mental storerooms rather than let them atrophy. A doctor from New York believed that more test takers would have passed had they devoted time to book reading rather than ball games, moving pictures, the sporting news, and other preoccupations of American males. Readers who followed the story flexed their cerebral muscles by taking the exam themselves. Men congregated in subways, clubrooms, college dorms, and hotel lobbies, jotting down answers to questions they speculated had been on the exam. People were wholly invested in establishing whether or not the exam tested scientific competence, and those with and without college training were curious about how they would perform on it if it did.[3]

As erudite as Edison appeared through all this, people seemed to forget that he had become who he was without the assistance of professional degrees of any kind. He never went to college; as a boy he was homeschooled and thrown quickly into business ventures to fend for himself. He observed the world around him and learned through reading and hands-on experimentation. As an established inventor he still boasted a subscription list of sixty-two periodicals, most of them scientific but also economic and legal and others oddly eclectic. Science and technology fascinated him, but so did geography, literature, and music—realms of knowledge that academic specialists considered "generalized trivia" in the technological age.[4]

Edison's hands-on experience of science reinforced his opinion that academic specialization, the hallmark of the modern university, had stifled human curiosity and compartmentalized men's thoughts until they knew lots about minutiae and nothing about anything else. He doubted that a modern college man could come close to filing a patent portfolio as large and varied as his, for it was unlikely that he was as inquisitive about the natural world. He put college men on the defensive at a time when they had sought authoritative status as experts. His exam had burst open a pregnant debate about college versus practical industrial training, but it also brought to the surface questions about the intellectual equipment of educated American men. In determining that his applicants were ill equipped to handle modern-day problems, Edison had essentially emasculated them.

Women observed these debates from the periphery and had their own opinions. "It is the men to whom we are accustomed to look for intellectual guidance who say the test could only be met by an 'encyclopedic mind,'" reflected Ellen Lynch of New York. She thought it "a matter of unsentimental fact" that there was "not a question in the list that could not be answered correctly by any well-trained boy or girl of 16." Who are the real best and brightest, another woman posited: masters of some atomized niche or broad thinkers with the capacity to draw knowledge from many parts of the social and natural world? Female readers of the *Times* seemed to relish the opportunity to respond sanctimoniously to these questions. Edison's test may have been of deeper significance to them than their husbands and fathers supposed, for it challenged the assumption that academic channels open almost only to men were requisites to a scientific mind. Perhaps women's hands-on study of nature—unpaid, without title, without resources, and outside the university—was of value after all. For a long time women had been told otherwise.[5]

The notion that women and science didn't mix had deep roots, traceable to Aristotelian and Platonic ideas about nature and knowing and Enlightenment views that supposed men and women's inherent traits to be complementary and oppositional. The Greeks linked femaleness with passive, indeterminate matter (and maleness with active, determinate forms), and in later generations the observation of fertilization under microscopes (interpreted as a motile sperm penetrating a passive egg) seemed affirmation. Eventually Kant, Rousseau, and other liberal philosophers (male by definition in this tradition) insisted that women were anti-intellectual by anatomy and default. They were lovers and feelers; men, thinkers and doers. Early positivists such as Newton, Descartes, and Locke also conceptualized links between external

disinterestedness and male knowing and internal subjectivity and female knowing. They grew confident that Nature was knowable through
the senses or experiment; but only men, not women, were equipped to
unlock its mysteries. Philosophers from Bacon to Goethe to Nietzsche
all assumed that the prerogative of Rational Man was dominance over
female nature; the mastery of nature and woman was essentially one
and the same.[6]

As the ideas of the scientific revolution took hold, they fostered the
rise of industrial capitalism, which in turn privatized family households.
Poor and working-class women and men went to work in factories,
though among the burgeoning industrial middle class the divisions of
labor and spheres of influence were more clearly distinguishable between the sexes. Ideal men occupied the public realms of learning, politics, paid work, and eventually professional science, while ideal women
occupied the supposedly subjective and sentimental domestic sphere.
Women were said to excel in such activities as cooking, cleaning, nurturing, and soothing daily quarrels; good science, as mirrored in the male
professional, however, required unemotional and empirical thinking in
abstract and universal terms. Whether or not individual women—or
men—possessed traits to support such binary views was irrelevant, for
cultural stereotypes possessed logic of their own.[7]

In the eighteenth century, a knowledge of "natural history" had
become part of the intellectual equipment of well-to-do women, and
it continued to be a popular subject of the lyceum circuit through the
nineteenth century. Botany in particular had long been cast as an avocation of women with leisure and means sufficient to cultivate herb gar-
dens and collect specimens for men to analyze. Editors of such domestic
periodicals as *Godey's* and *The Ladies Repository* encouraged women to
attend public lectures that would help them to nurture their children's
love of plants, assuming that women readers were sentimentally drawn
to the care of living things. Mrs. C. M. Badger illustrated lithographs of
flowers and shrubs, requiring painstaking classification of genus and
species. In the taxonomy of professional men, however, her books were
classified as nonscience; they were creative endeavors inspired by a female love of beauty rather than by a penchant for scientific truth.

Jane Marcet was among the women who wrote best-selling science
books in the nineteenth century—astonishingly, in the field of chemistry, not botany. Her success can be ascribed to the appropriately ped-

agogical tone and style of her work, often organized as conversations between mother and child. Her timing, too, was critical to her success, since she published before she could become associated with the concepts of amateurism that stigmatized women later in the century. By the end of the nineteenth century, the woman scientist had become the ultimate oxymoron; one science editor likened a woman's fondness for experiment to a perversion that needed to be concealed. Normal women were nurturing and noncompetitive, but modern science was imagined as the opposite. Women like Marcet were acceptable because they wrote watered-down science for children, busying themselves with the common work of popularization to free men to conduct prestigious research.[8]

Professional men accepted women curators and collectors, textbook writers, and illustrators, since such work dovetailed neatly into the work of women custodians, pedagogues, and helpmeets. Anna Botsford Comstock drew flora and fauna for the textbooks of her husband and eventually became one of the most recognizable science writers for children and lay people in the United States. American schoolteachers referred to her *Handbook of Nature Study* as the "Nature Bible," and she became so influential a lecturer and writer that the League of Women Voters chose her as one of the twelve greatest women in the United States. But her greatness was rarely couched in terms of science. In the minds of men, the pedagogue, popularizer, editor, and engraver of natural history was literary and sentimental, at best an amateur.[9]

The associations between amateur science and womanhood rigidified as men enjoyed more government funding in public universities to which women had little access. The American Association for the Advancement of Science divided itself into specialized societies of men, each of which established specialized journals and agendas. And yet discussion about women's scientific abilities in popular journals such as *Scientific American* and *Popular Science Monthly* indicated that there was no total consensus on the issue. After the Civil War, women's colleges, among them Mount Holyoke, Wellesley, and Barnard, offered science courses in particular fields that aspired to the rigorous standards of male institutions. By the 1890s some of the nation's most prestigious universities—Cornell, Chicago, Berkeley, and others—accepted women students, several of whom aspired to be scientists. Harvard medical professor Edward Clarke insisted that women engaged in such pursuits to the detriment of their reproductive health, but social scientists, many of them women, made a more compelling argument for the environmental underpinnings of gender difference than biological rationales.[10]

With degrees in hand, modest numbers of women began to enter scientific institutions that were originally bastions of men—albeit in marginalized capacities as librarians, computers, secretaries, and assistants. They had the academic credentials but not the titles and salaries of scientists. They were segregated within fields but also across fields—horizontally as well as vertically. Women were ejected from fields deemed "hard" in the name of higher standards—essentially a strategy of containment. Professionalization continued to be a gendered and gendering process, one designed to ameliorate *softening* by reinstating masculine status in some fields over others.[11]

Several meanings are implicit in the metaphorical use of "hard" and "soft": Hard science came to be understood as intellectually and physically rigorous, fortified by airtight, indestructible positivist methodology. Subjects of "hard" research were physically inanimate, not squishy, oozing, or alive; results of "hard" research were consequently "hard and fast," versus the ambiguous findings of social scientists or other increasingly female investigators. Hardness and softness had long been assigned gendered connotations (the penetrating mind and hard muscles of men and the sentimental thinking and soft curves of women), and now through associative logic men funneled women into "softer" fields of lesser prestige. At some universities, including Yale, New York University (NYU), Cornell, MIT, and the University of Chicago, women instructors filled niches of expertise in hygiene; nutrition; and "social," "physiological," and "domestic" science.[12]

One of the pioneers of female science in the late nineteenth century was Ellen Swallow Richards, a woman who had worked with astronomer Maria Mitchell as an undergraduate at Vassar. When she couldn't find a job after college she decided to study chemistry at MIT, where she was admitted as a "special student" so that her name wouldn't appear on the university roster. She sewed buttons and swept floors to gain men's acceptance in the lab. For decades the university used her services as an instructor of chemistry and engineering, most of the time without compensation or title. As much as her story sounds like a victimology, one can also see her as an agent of her own destiny, a woman who used domesticity as camouflage to become master of an uncontested domain: the "Woman's Laboratory," where she quietly innovated in the fields of ecology and sanitary engineering. Rather than rue the day she was relegated to domestic science, Richards established organizations and scholarships in her separate and distinctly female expertise. Her investigations into ventilation, nutrition, and clean water were of great social significance, even if their importance wasn't reflected in her promotion or pay.[13]

Richards's strategy made sense, given the climate of institutional science, but it begged the question: In separateness, had she achieved professional equality? Younger women thought not. In earlier years they could rely on husbands and fathers to get them access to labs, but after 1910 women scientists, who were largely relegated to women's colleges and women's fields, felt more removed from the scientific establishment then ever before. In politics young women increasingly rejected the separatist tactics of their mothers, who wanted to win suffrage on the basis of their special moral authority as women; in the lab these daughters wanted to compete and interact directly with men, not as women, but as scientists. They were heartened when the U.S. Civil Service Commission decided to administer its qualifying examinations on a sex-blind basis in 1921. For the first time they could be hired, in theory, as researchers for agencies such as the Federal Bureau of Standards. In practice the hiring for government jobs was hardly sex blind, however, and at colleges and universities, where science departments made similar claims, the results were equally ambiguous. Although women experienced a slight increase in hiring at coed college campuses after 1920, when they were 20 percent of all employed women scientists, the promise for more equitable hiring practices brought about a reverse trend: the women's colleges that women once relied on for employment sought out more men to lend prestige to their programs. Earlier in the century 58 percent of women scientists worked at women's colleges; by 1920 the total had decreased to 37 percent. The percentages were not balanced by increased numbers of women hired by coed institutions.[14]

Some women looked outside the university to industry, for World War I had expanded photographic, communication, and ballistic technologies, among others. Marie Curie, the most acclaimed woman scientist in the world, left her lab for the battlefront during the war; her vehicle, "the little Curie," carried X-ray equipment to makeshift medical facilities on the frontlines, proving that women, too, could provide significant resources to the allegedly masculine realms of science and war. In the United States, women who would have been shunted into home economics were producing gas and projectile weapons, explosives, instrumentation, and other war materials. A select few, such as Wellesley graduate Louise McDowell and Vassar graduate Frances Wick, worked on radar at the National Bureau of Standards and on airplane radios for the U.S. Army's Signal Corp, but their access to research was short lived. After 1918 women were laid off and men reinstated.[15]

By 1921 women's advances in science were still modest, especially since they enjoyed more access to science degrees than employment opportunities in science fields. Yet even with their small numbers, men feared a crisis of impending feminization in fields they had taken pains to infuse with power and prestige. Edison unwittingly proved that the crisis had come to a head. By doubting the qualifications of college-educated men, he was also doubting the dominant portraits of American manhood and men's innate scientific talent, bringing these issues to the surface, where men and women could weigh in.

The very day the *Times* first reported on Edison's exam, it also ran a story on Marie Curie's arrival in New York, the beginning of a six-week tour of the country. She was coming to collect a gram of radium donated by American women and would soon find herself at the center of tensions between the old and the new science, between feminists and the male establishment, and between Americans who differed on men's and women's proper spheres of influence. Men used her tour to showcase American science and industry, while educated women made it their coming out of sorts—a chance to make the case for their place in science and professional life. Curie was their living proof of success.

Notes

1. Robert Kargon, *The Maturing of American Science* (Washington, DC: American Association for the Advancement of Science, 1974), 2–3, 33; "Edison Asserts He Wants College Men If He Can Get Them," *New York Times*, May 16, 1921, 1. Hereafter, *New York Times* cited as *NYT*.

2. "Controversy Rages over Edison Test," *NYT*, May 15, 1921, 14; "Edison Questions Stir up a Storm," *NYT*, May 11, 1921, 6; "Edison Dashes Off New Questionnaire," *NYT*, May 14, 1921, 9.

3. "Edison's Questions Still Puzzle City," *NYT*, May 12, 1921, 17; "Here is Edison's 4-Column Sheaf of Knowledge," *NYT*, May 13, 1921, 10; "Edison Is Upheld by Many Writers," *NYT*, May 19, 1921, 20; "Edison Brainmeter Divides the Critics," *NYT*, May 17, 1921, 8.

4. See details of Edison's early years, in Randall Stross, *The Wizard of Menlo Park: How Thomas Alva Edison Invented the Modern World* (New York: Crown, 2007); Gene Adair, *Thomas Alva Edison: Inventing the Electric Age* (New York: Oxford University Press, 1996); Paul Israel, *Edison: A Life of Invention* (New York: John Wiley and Sons, 1998); "Edison Dashes Off New Questionnaire."

5. "Edison Brainmeter Divides the Critics."

6. Elizabeth Fee, "Women's Nature and Scientific Objectivity," in *Woman's Nature: Rationalizations of Inequality*, ed. M. Lowe and R. Hubbard (New York: Pergamon Press, 1981), 9–22; Genevieve Lloyd, "Reason, Science, and the Domination of Matter," in *Feminism and Science*, ed. Evelyn Fox Keller and Helen E. Longino (New York: Oxford University Press, 1996), 41–52; Carolyn Merchant, "Isis' Consciousness Raised," in *History of Women in the Sciences: Readings from Isis*, ed. Sally Gregory Kohlstedt (Chicago: University of Chicago Press, 1999), 11–15; Lisbet Koerner, "Goethe's Botany: Lessons of a Feminine Science," *Isis* 84 (1993), 470–95; Margaret Wertheim, *Pythagoras' Trousers: God, Physics, and the Gender Wars* (New York: W. W. Norton, 1997), 148–54; Evelyn Fox Keller, *Reflections on Gender and Science* (New Haven, CT: Yale University Press, 1995), 19, 31–44.

7. Sandra Harding, *The Science Question in Feminism* (Ithaca: Cornell University Press, 1986), 60–64, 123–25; Londa Schiebinger, *Has Feminism Changed Science?* (Cambridge, MA: Harvard University Press, 1999), 70–71.

8. Sally Gregory Kohlstedt, "In from the Periphery: American Women in Science, 1830–1880," *Signs* 4 (Autumn 1978), 81–96; Deborah Jane Warner, "Science Education for Women in Antebellum America," in *History of Women in the Sciences*, 193; M. Susan Lindee, "The American Career of Jane Marcet's *Conversations on Chemistry, 1806–1953*," *Isis* 84 (1993), 470–95; Barbara T. Gates and Ann B. Shteir, eds., *Natural Eloquence: Women Reinscribe Science* (Madison, WI: University of Wisconsin Press, 1997), 3, 11–17.

9. Comstock was in fact an "academic," who taught at Cornell University for decades before being tenured at the age of sixty-five. Pamela M. Henson, "'Through Books to Nature:' Anna Botsford Comstock and the Nature Study Movement," in *Natural Eloquence*, 116–39; Marcia Myers Bonta, *Women in the Field: America's Pioneering Women Naturalists* (College Station, TX: Texas A&M University Press, 1991), 152–64.

10. Sally Gregory Kohlstedt, "Parlors, Primers, and Public Schooling: Education for Science in Nineteenth-Century America," *Isis* 81 (1990), 425–45; Kohlstedt, "In from the Periphery," 90–91; Rosalind Rosenberg, *Beyond Separate Spheres: Intellectual Roots of Modern Feminism* (New Haven, CT: Yale University Press, 1982), xv, 1–27. Studies of women social scientists include Helen Thompson Woolley, *The Mental Traits of Sex: An Experimental Investigation of the Normal Mind in Men and Women* (Chicago: University of Chicago Press, 1903); Leta Stetter Hollingworth, "Comparison of the Sexes in Mental Traits," *Psychological Bulletin* 25 (December 1918), 427–32; Elsie Clews Parsons, *The Old-Fashioned Woman: Primitive Fancies About the Sex* (New York: Putnam, 1913).

11. Margaret Rossiter, *Women Scientists in America: Struggles and Strategies to 1940* (Baltimore: Johns Hopkins University Press, 1982), xvi, 29–72, 98–99, 107; Harding, *Science Question in Feminism*, 62.

12. Schiebinger, *Has Feminism Changed Science?*, 160–64.

13. Sarah Stage, "Ellen Richards and the Social Significance of the Home Economics Movement," in *Rethinking Home Economics: Women and the History of a Profession*, ed. Sarah Stage and Virginia B. Vicenti (Ithaca, NY: Cornell University Press, 1997), 21–22.

14. Daniel Kevles, *The Physicists: The History of a Scientific Community in Modern America* (Cambridge, MA: Harvard University Press, 1987), 202–3.

15. Rossiter, *Women Scientists in America*, 116–18.

1 Madame Curie's American Tours: Women and Science in the 1920s

IT WAS AN INTERVIEW TO RELISH: THE RECLUSIVE MADAME CURIE agreed to sit with American editor Marie Mattingly Meloney for a profile to be published in Meloney's magazine, the *Delineator*. Stéphane Lauzanne, editor-in-chief of *Le Matin*, had been following the Nobel laureate's story for years. "She will see no one," he warned Meloney. "She cannot understand why scientists, rather than science, should be discussed in the press." For all her attempts to prevent it, Curie's had become a household name. She had not made a public appearance in fifteen years, and yet she agreed to meet this American journalist in 1920.

Meloney had met famous scientists before. She had grown up near the estate of the illustrious Alexander Graham Bell, a breeder of horses that she yearned to ride. And only weeks before arriving in Paris she had stood in the laboratories of Thomas Alva Edison; eyeing the new-fangled equipment at his command, she decided that one's scientific prowess brought not only admiration, but also great financial reward. In Pittsburgh she had seen the smokestacks of the greatest radium-reduction plants in the world. And yet when she reached the physics building at the rue Pierre Curie, the originator of this technology was a pale, timid woman, surrounded by nothing that would suggest material gain for her efforts. Her office was sparse; even to an untrained eye, the facilities looked inadequate. Curie rubbed the tips of her fingers over and over the pad of her thumb, a habit she had developed while trying to regain feeling lost in her hands. For her, scientific discovery had not led to riches but physical and material sacrifice.[1]

That the two women would hit it off famously could not have been predicted, for Meloney was in many ways Curie's younger antithesis. An important person in the U.S. editing world and in New York society, she had started out as a reporter for the *Washington Post* but had become one of the most influential women of American media. She exuded social grace, money, and a penchant for publicity. Curie, a Pole who had emigrated to France to pursue science, was not interested in appearances, including her own. She dressed plainly, typically in black, and preferred to be undisturbed by anyone, at home or in the lab. Male colleagues she met at professional meetings thought her terse and disagreeable; small talk was not her métier. Her closest confidant since her husband's death in 1906 was her daughter Irene, a young physicist who was like her mother in interest and temperament. Despite their differences, however, Curie and Meloney were also working mothers who understood each other's struggle to balance professional and family obligations. The interview at the Sorbonne lengthened into informal talks during Curie's personal time. Meloney became "Missy" and Curie "Marie"; a mutual affection took root that lasted the rest of their lives.

Meloney felt confident that her readers wanted to hear from the reclusive scientist, for Americans had been following her story for years. In 1904, *Vanity Fair* had presented Marie and Pierre as a couple, months after they had together won the Nobel Prize for radioactivity. Marie appeared to be a puzzling contradiction: a brilliant scientist who was, nevertheless, also the "woman behind the man" in the Curie partnership. Although she was the codiscoverer of radium, she was also the person responsible for family domesticity, an image confirmed by another woman journalist who covered "The Curies at Home" for *The World To-Day* the same year.[2] The domestic image may have seemed to diminish her as a scientist; yet young women with career aspirations saw her differently, as a model of grace and competence outside the home. When Curie won her second Nobel Prize in 1911, science editor James McKeen Cattell cast her in *Science* and *Popular Science Monthly* in an overtly feminist light, calling her rejection from the *Académie des Sciences* a tragedy for women scientists and science alike.[3]

The journalistic treatment of Curie was startling, given the paucity of coverage on women scientists generally in the American press. Taking into account scientists both as authors and biographical subjects, male scientists were fifteen times more visible than their female counterparts in mainstream magazines. Journalists described science as an endeavor requiring culturally virile attributes—emotional detachment, intellectual objectivity, even physical strength at times. In 1920 a Columbia scientist described the "Eminent Chemists" of his day as the "the Dickenses," "the Thackerays," and the "Wells" of the field; then he heralded Curie as "the Columbus who discovered another continent in science," feigning amnesia about her gender to write about her in grandiose terms.[4]

Still, even as Meloney talked with Curie, American women stood on the brink of winning suffrage. It was the increasing reality that domestic women were making inroads into political, professional, and public endeavors. Meloney thought that writing of Curie's achievements might allow other science-minded women to gain social acceptance. And who better than Curie to inspire an educated readership? She held a doctorate degree, had won Nobel Prizes, and had proved that, despite the odds, a woman could reach the highest echelons of her chosen professional field—even while rearing two fatherless daughters. She was a single mother and a world-class scientist—an everyday woman who made greatness look reachable. Meloney thought American women would see her as an icon for achieving it all: marriage, family, and career.

But as she told her story, Curie's frustrations became apparent. French science was impoverished after World War I, making it impossible to procure radium, the very element she had discovered, and as a result, her research had been curtailed. She could account for the locations of some fifty grams of radium in the United States; four were in Baltimore, six in Denver, seven in New York.

"And in France?" Meloney queried.

"My laboratory has hardly more than a gram," Curie replied, and that bit was available only for extracting emanations for cancer treatment in hospitals.

Meloney begged for clarification: "*You* have only a gram?"

"I? Oh, I have none. It belongs to my laboratory."

"But surely revenues from the patent of radium can pay for more," Meloney suggested.

"There were no patents," Curie corrected. "We were working in the interest of science. Radium is an element. It belongs to all people."[5]

Curie thought it wrong to profit from her discovery, and yet the discovery had most assuredly lined the pockets of chemical producers back in the United States. American companies had mastered her processes to become the foremost producers of her coveted radium. Standard Chemical Company of Pittsburgh was responsible for more than half the 140 grams in existence worldwide, but it wasn't cheap. Medical researchers bought it in milligram increments for $120 apiece, and Curie needed more—at least a gram if she could get it. Meloney listened intently, the wheels turning in her mind. She prided herself on making things happen; she was going to get Curie her radium.[6]

A public relations strategy developed in her imagination in the form of a story that would play out on the pages of the *Delineator* and major publications throughout the United States. It had all the elements of an American legend in the making: a tale of travesty about a woman and mother, who had sacrificed for the world but had received nothing in return. This woman had endured abject poverty to discover the element that was the leading cure for cancer, but had refused to patent its production so that the researchers of the world could study it readily. She should have been heralded as a saint, and yet the price of her altruism was a debilitating lack of funds, not for her own creature comforts but for humanity, who continued to suffer without her life-saving research. "I had been prepared to meet a woman of the world, enriched by her own efforts and established in one of the white palaces of the Champs d'Elysées or some other beautiful boulevard of Paris," Meloney wrote. "But Madame Curie was a simple woman, working in an inadequate laboratory and living in an inexpensive apartment, on the meager pay of a French professor."[7]

The synopsis seemed extreme to Curie, who insisted that French scientists worked under modest conditions all the time, but Meloney wouldn't hear it. In the United States, she told Curie, many women had money, and when she returned from Paris she would write letters to the wives of physicians in major medical societies. Meloney planned a broad campaign that would bring together an odd mix: cancer researchers, academic scientists, chemical companies, and women with money and social connections. American science had grown quickly after World War I so that government funding was no longer enough to maintain it, and the age of fund-raising in the name of "science as social cause" had emerged. Meloney hoped to exploit Curie's image just as chemical companies were exploiting Albert Einstein's to raise funds for themselves and his educational initiatives overseas.[8]

But the campaign could only succeed if donors perceived the scientist and her work as forces of benevolence. When Alice Hamilton and Ellen Swallow Richards were praised in the press for the healing applications of their work, American women who wanted to emulate them entered nursing and medical fields in larger numbers than ever before. But they studied organic material rather than the lifeless stuff of physical science. Meloney's promotional strategy necessarily had to reinvent Curie's physics, turning inanimate radium into a regenerative force for healing—and reenvision Curie's life, turning her into a benevolent lady. The press coverage on medical radium strengthened Meloney's view of how to proceed. The *New York Times* reported more than 150 deaths directly linked to cancer and radium research, its victims including men who had once worked under the Curies themselves. As more of the world's radium experts succumbed to mysterious symptoms, Curie's commitment to research seemed all the more heroic in Meloney's hands. The burns on her fingers, the jaundice, and worsening cataracts made Curie a martyr to behold; the findings of her research would be her gift to the world.[9]

Curie had other compelling reasons to team up with American sponsors. During World War I she had been charmed by the simple demeanor of the young American soldiers who came to Paris to study radioactivity. She was also grateful to Andrew Carnegie, who had established a fund to keep her lab staffed after the death of Pierre. Over the years American women had expressed their admiration for her work, but she still couldn't imagine what would compel them to write checks—or in most cases, to have their husbands write checks—so that a French scientist could pursue her research. Meloney explained that they would see a mutually beneficial arrangement before them: as they raised money for her radium, they would exploit her image for their humanitarian and professional ends. This was false advertising, Curie told her. She had abstained from ideological movements her whole life to work in the spirit of "pure science" and wanted nothing to do with developing medical applications for radium or advancing the cause of women. Don't taint my radium with agendas, she implored; it absolutely "must belong to science."[10]

A perfect strategy in the eyes of a privileged career woman like Meloney was apparently not acceptable to a scientist like Curie. To base a campaign on her maternal motivations was to make her look emotional, subjective, and hence defective to male scientists. The maternalism that women evoked to win access to college and the vote was antithetical to the idea of science that Curie had internalized and come to revere. But for the sake of obtaining her radium she deferred to Meloney—it was all about the science in the end.

Against her better judgment Curie began writing a biographical piece to be distributed in the United States. Meanwhile, Meloney called on her connections in the world of publishing to get dozens of articles into print during the coming months. *Current History Magazine* printed "The Story of Radium in America." Dr. William Mayo asked *Delineator* readers, "Do You Fear Cancer?" drawing out anxieties that supporting Curie's work would presumably allay.[11] Although these pieces were not explicitly about Curie, they set the stage for solicitations on her behalf by dramatizing the dangers of cancer and its study. Articles that followed emphasized Curie's importance to the field and shaped images of her that were appealing enough to make Americans open their hearts and checkbooks for the medical breakthroughs she ostensibly represented. Editors for *Century Magazine* and *Current Opinion* were hardly subtle, calling Curie a Jeanne D'Arc of the laboratory and "the most famous woman in the world." Meloney praised her in print for "minister[ing] to an agonized people." The front page of her magazine declared Curie's raison d'être: "That Millions Shall Not Die!" Her face, Meloney wrote, was "softer, fuller, more human" than most. "It had suffering and patience in it," as did "every line of her slender body." In its totality, hers was "the mother look" epitomized.[12]

Meloney's publicity campaign was designed to appeal to a wide swath of people. Modern career women would admire Curie's professional competence, ambitious young people her self-making; she had achieved the American dream, even if not on American soil, pulling herself up by the proverbial bootstraps to work tirelessly for success. Unlike the aristocratic scientists of Europe, she apparently never forgot her modest Polish roots. It was not beneath her, Meloney told readers, to launder her own undergarments, even when she was hosted in homes that had servants. She was the embodiment of the Puritan ethic, much as Americans fancied themselves.[13] Too poor to pursue science formally, she had borrowed books from the factory library and earned money as a governess to purchase a fourth-class ticket from Poland to Paris. With almost nothing but the clothes on her back, the twenty-four-year-old had arrived in the Latin Quarter, where she continued to exercise the frugality of legend. Journalists and biographers sensationalized her student years in a sixth-floor walk-up. Allegedly she hauled her own wood for a furnace that didn't work and resorted to piling clothes on the bedcovers and anchoring furniture on top to create a shield from drafts. Some maintained that she treated herself to an occasional boiled egg or piece of chocolate; others insisted she lived on tea and crusts of bread. All, including Curie herself, described a young woman so fixated on her studies that she didn't think to eat until her body collapsed. Her notebooks were obsessively neat. She graduated first in physics at the Sorbonne in 1893 and second in mathematics a year later.[14]

Curie had many professional and social acquaintances, yet biographers insisted that the only person who caught her attention in those monastic years was Pierre, a man of equal intensity. A Polish professor introduced her to the thirty-five-year-old physics professor at L'École Municipale in 1894. In addition to working on crystals and magnetism, he and his brother had inauspiciously discovered piezoelectricity and were using their findings to design instruments, including the electrometer Curie used to confirm the existence of radium. The first gift Pierre offered his future wife was a paper on the symmetry of electric and magnetic fields. He must have sensed their unique connection, for there was no better gesture of affection in her eyes. They were kindred spirits; neither ate, slept, or remembered social niceties when in the throes of some scientific question. Pierre loved Marie's commitment to science. Women of her intellect and drive were so rare, he thought, that he had to have her for himself. Their single-minded devotion to science should be their gift to humanity, he told her, and she eventually agreed. The path they walked together they described as an "anti-natural" one: they renounced "the pleasures in life" and pared down all distractions outside their immediate family to live and breathe their research. This meant renting a modest home, eating and dressing plainly, and rarely entertaining guests. "All preoccupations with worldly life were excluded from our existence," Marie recalled fondly. In a scientific man such reclusiveness was expected, but Marie loved Pierre because he accepted the same in her.[15]

Their wedding in 1895 was simple; there was no exchange of rings or festivities to follow. Marie chose a navy blue dress versatile enough to wear at the wedding and in the lab afterward. She and Pierre enjoyed a bicycling honeymoon in the French countryside, but they didn't stay away from the lab for long. Pierre resumed his formal position at L'École

Municipale, while Marie continued her studies and served as his assistant gratis. Marie agreed to be the homemaker so long as the work was confined to rudimentary tasks. Aside from used furniture relatives had given them, she had little to keep clean. She was not talented as a cook, but Pierre seemed not to notice what she was feeding him anyway. Their three-room apartment was on the rue de la Glacière, in close proximity to the school of physics so they could spend more time in the lab.[16]

Marie had been investigating the magnetization of tempered steel when recent findings caught her attention. On the heels of Wilhelm Röntgen's discovery of the X-ray, Henri Becquerel, who had been studying uranium salts, had noticed that they emitted special rays. Marie made it her mission to measure them and discern the conditions under which they existed. She confirmed that they were in fact an atomic property of uranium and that samples of thorium behaved similarly, but the rays emitted in her samples of minerals were not commensurate with the amounts of uranium and thorium she knew to be in them. Her observations led her to a theory: the ores contained trace amounts of another substance, one more radioactive than any known element. After grueling experiments Marie discovered that in fact two new substances were involved. She identified one in 1898 and named it polonium, after her homeland. Five months later, she identified the second element, which came to be known as radium.

The problem then became isolating the elements. Pierre was studying crystals but found himself increasingly swept up into his wife's doctoral research. While he studied the properties of radium, Marie extracted pure radium salts. The Austrian government had disposed of tons of pitchblende, a uranium ore, which Marie had obtained cheaply in order to begin the painstaking process of purification. She had to dissolve the dusty substance in acid before chemically separating its elements, which meant transferring vast vessels of liquid and precipitate and, with an iron bar, stirring cauldron after cauldron of the substance. The tasks were laborious; they left her exhausted and she fell ill with pneumonia. She extracted thousands of gallons of distilled liquid from her mounds of pitchblende, ultimately yielding amounts of radium solution that barely filled a thimble.[17]

The work wasn't glamorous, but it had been romanticized in the American press. Readers were transfixed by the image of Pierre and

Marie conducting experiments side by side in their dilapidated shed. Inside they performed the bench work; in the adjoining yard, the bulkier transport of pitchblende. The building's windows were broken and its heating unit didn't work; at the height of their experiments Marie logged an indoor temperature of just above freezing. They worked for hours in silence, stopping only to put on sweaters or warm themselves with cups of tea made at a rickety stove. There was one detail publicists made more of than Curie herself: amid the hauling, stirring, measuring, and recording she had in fact given birth to a child. Taking none of the standard precautions of modern nuclear labs, Marie fed Pierre lunch probably laced with radioactive material. She took in harmful fumes herself, which contributed to her chronic illnesses and her frustrations with nursing.[18]

Still, without qualification Curie recalled the years in the shed as the happiest of her life—not because of newfound motherhood, but rather because of her unfettered access to scientific work. The bliss only ended when her discovery of radium brought unwanted celebrity, making the "anti-natural" ideal impossible to maintain. She was entirely in charge of her baby's care and she was also in the midst of doctoral research. Desperate for domestic help, she searched for wet nurses, housekeepers, and nannies, which also took her away from the lab, as did the additional teaching hours she had to take on to cover the cost of her helpers. She secured a position at a girls' school in Sèvres, well outside Paris; the commute was yet another diversion from the experiments she was desperate to run.[19]

Curie's father-in-law, a widower, came to her aid and helped to care for baby Irene, as his own. The balancing act became easier, but even then, Marie thought the baby a distraction—a perspective that American biographers then treated ambivalently, as others have since. Some have insisted that she doted on her daughter—recording the baby's first teeth, steps, and words and making her clothes by hand. Indeed some personal notebooks verify these details, but they also read like lab notes, clinical and joyless, for what could not be measured or quantified precisely she simply left unrecorded.[20] Curie admitted privately that the responsibilities of motherhood kept her from the work she loved the most. The night she isolated her pure radium chloride, she and Pierre left Irene sleeping in her crib to return rain-soaked to the lab. Like doting parents they checked in on their other sleeping baby—a test tube of radium that filled them with pride when it started to glow. Unambiguously, Curie described radioactivity to one of her students as "the child to whom she had given birth"; she would nurture it and devote to it "the whole of her working life."[21]

Physicist George Sagnac expressed concern that Marie's unbalanced family life bordered on obsession and neglect; he worried about her health and her daughter's: "Don't you love Irene? It seems to me that I wouldn't prefer the idea of reading a paper by Rutherford to getting what my body needs and looking after such an agreeable little girl." Pierre refused to be critical; ultimately the baby was his wife's unexpected burden more than his. He was given a full appointment at the Sorbonne, as Marie struggled to balance research, teaching, domestic duties, and child care. This was not the dream he had promised her. Marie resented her obligations, but never Pierre: "He used to say that he had got a wife made expressly for him to share all his preoccupations. Neither of us would contemplate abandoning what was so precious to both."[22]

The world knew 1903 as the year the Curies won their Nobel Prize. For Marie, it was the year she lost her privacy and a second child. In August she prematurely delivered a baby girl who died hours later. Letters to her sister revealed how angry she was at herself for thinking her body up to the rigor she had put it through leading up to the birth. She had continued the taxing purification of pitchblende, stopping only to take the bicycle holiday that had induced her early labor. She felt more optimistic the following year when she gave birth to a healthy daughter named Eve. She recalled the added burden lightheartedly in the biographical piece she wrote for Meloney: "It was not easy to reconcile these household duties with my scientific work, yet, with good will, I managed it."[23]

If she sounded like the model of contented motherhood, the reality was more complicated. It was Grand Pé who took Irene to see the sites of Paris and the nanny who taught the girls her native Polish. For whole summers the girls stayed with hired help in Brittany so that Curie could work uninterrupted back in Paris. Irene wrote letters to her dear "Mé," imploring her to see them, and felt devastated when she received in return only one letter of apology. Eve, too, felt cheated, that she could never compete with her mother's science. While Irene worked harder to become the lab partner with whom her mother wanted to engage,

Eve turned elsewhere, becoming a musician on the international stage. Marie and Irene avoided publicity, but Eve loved to be seen. She had a penchant for makeup and heels and contemplated becoming a journalist; her mother thought she might just as well become the devil himself. Only in later years did the bond between them strengthen, especially when Eve nursed her mother through her last illness. After Curie's death, Eve became committed to political and social causes—taking the antithesis of her parents' anti-natural path.[24]

In emphasizing the early years of Marie's marriage and career, Americans sentimentalized her as a mother and whitewashed the chauvinism that pervaded her career before and after she had children. No one mentioned the stares of male students and faculty, who thought her an interloper on their sacred ground. Pregnancy and motherhood had divided her energies in ways male colleagues couldn't appreciate. Even in the most liberal of French circles, her return to work so soon after birthing seemed to border on neglect. Men wanted to believe that her extracurricular activities (read: domestic) gave them cause not to take her seriously, but they had to; she performed better than any of them on the Concours d'Agrégation, and the dissertation that followed stunned her evaluators, future Nobel winners themselves. She was the first woman in Europe to complete a doctorate in her field, and she did it with force and style. Not only had she discovered radium, but she had isolated it and determined its atomic weight. The phenomenon of radioactivity was no longer a mystery—and yet a formal academic appointment for her remained elusive.[25]

The scientific community looked on in disbelief when she and Pierre won the Nobel in physics with Becquerel in 1903. The distinction didn't come easily; the French Académie quietly lobbied for the prize to be split between Becquerel and Pierre only, a ploy a Swedish physicist made known to Pierre. No prize for radioactivity could possibly exclude his wife, he implored. *She* first suspected the existence of the elements they discovered, and *she* had already run experiments when he entered the fray. Upon conferring the prize, the president of the Swedish Academy of Sciences cited the old proverb "Union is strength," acknowledging the Curies as a team. He also thought it apropos to cite a biblical passage: "It is not good that man should be alone; I will make a helpmeet for him." His assumption that Marie served the team as helpmeet was pervasive in the science community. Most believed she was an assistant rather than a partner, a doer of deeds that Pierre brilliantly conceived. For now Pierre's assertions shielded her, but they did not change what men thought in the back of their minds.[26]

British physicist Hertha Ayrton, Curie's trusted confidant, suffered similar suspicious stares from men who assumed her successes to be her husband's. "Errors are notoriously hard to kill," she asserted, "but an error that ascribes to a man what was actually the work of a woman has more lives than a cat." Curie stayed close to Pierre to enjoy the refuge from scrutiny, but she also created strategic distance, publishing her own papers that demonstrated her ability to conceptualize and perform original research.[27]

She found her reputation harder to defend after Pierre was tragically killed in a traffic accident in the rue de Dauphine in 1906. American publicists insisted that devotion to her children saved her from deep despair, but she said it was the promise she and Pierre had made to each other that kept her focused and gave her purpose. She refused a widow's pension, taking over Pierre's academic appointment at the Sorbonne instead. University officials went along—even if they appointed her as an assistant rather than give her Pierre's full professorship.[28]

The following November, reporters, dignitaries, and such legendary figures as Lord Kelvin himself sat among the advanced students in the amphitheater in which Curie gave her dead husband's lectures. She walked in to booming applause that she tried to suppress with gestures for quiet. People wondered if she'd react with displays of emotion and odes to Pierre. Instead she proceeded in a steady monotone: "When we consider the progress made by the theories of radioactivity since the beginning of the Nineteenth Century . . ." Students were amazed; she had picked up on Pierre's last explication of polonium almost to the sentence. Even in the wake of Pierre's death, she refused to give colleagues reason to think her anything but the consummate professional. She had to convince the audience that she could act the part of the disinterested scientist as well as any man with whom she now competed.[29]

But when the period of mourning was over, reaction to her university appointment changed; invariably colleagues and the press began to look critically at the widow propelled by her husband's death. Pierre had been elected into the Académie des Sciences in 1905, but his colleagues took less kindly to his widow's submission as a candidate five years later.

A war of the sexes played out in the press, for the Académie had a long, prestigious history of being open only to men. Her detractors found it easier to build their case against her when they seized on allegations from the wife of physicist Paul Langevin that he and Curie were lovers. Langevin had been a former student of Pierre's and a close family friend, making the alleged affair all the more scandalous. Langevin's wife released supposed love letters to the tabloids, creating a publicity nightmare from which Curie never recovered.

If it all seemed unrelated to her science, members of the Académie refused to see it that way. Feminists defended her, but right-wing detractors scrutinized her fitness for membership on every score. In the illustrated daily *L'Excelsior*, a physiognomist and handwriting expert examined the published love letters to determine her qualifications. Nationalist publications used her foreign origins against her; she became the object of anti-Semitic slander, though she was not a Jew. A prominent French editor wrote of her in scathing terms: she was no longer the "Vestal Virgin of Radium" but a Polish interloper with inappropriate ambitions; first she rode the coattails of Pierre, and now she looked to his unsuspecting students.[30] Albert Einstein committed adultery and left his wife without ill effects in the press, but Curie was different. Although no longer bound in matrimony, she had mistakenly revealed herself as a sexual woman, and thus the fiction of her saintly science was impossible to maintain. In the midst of scandal, Curie also made headlines for winning her second Nobel Prize. It should have shored up her unquestioned status in science, but it only led to greater scrutiny. The Nobel Committee seemed to be splitting hairs, some suggested, awarding a prize for her chemical study of radium when the first one she won was for radioactivity in general. Curie accepted her prize reticently; Irene, now a fourteen-year-old, sensed the tension surrounding her mother in Stockholm. Marie credited Pierre but also with indignation referred to "*my* hypothesis about radioactivity," "*my* chemical work to isolate radium." Without Pierre there to defend her, it is unlikely she was convincing.[31]

The allegedly adulterous Langevin went back to work unscathed, but for Curie the sex scandal seemed to flick a switch; her trespass into the masculine realm of science could no longer be excused. The scientific establishment decided that she couldn't win individual accolades with her reputation intact. Already reclusive, Curie tried to evade the public eye at all costs. For the following decade she refused almost all requests for interviews, no matter the context, making it all the more curious that she agreed to talk with Meloney in 1920. Perhaps Curie hoped that the staunchly independent American would provide both a sympathetic ear and the power to advance her scientific work. The way to do it, Meloney insisted, was to reinstate her maternal image among Americans more willing to forgive her frailties. She had the word of every leading newspaper editor in New York that he would make no mention of the love affair during her Radium Campaign. Arthur Brisbane, managing editor of the *Evening Journal*, even handed over his file on the scandal to let Meloney do with it as she pleased.[32]

The Curie who was introduced to Americans appeared as the Virgin Mary in black—maternal, saintlike, in no way sexual. Her refusal to patent her method for extracting radium was presented as an absolute sign of her altruism, and her tragic circumstances as a widow only added to the image. Curie was to be the "common woman" in a country where the common man was beyond reproach. In press releases the title "devoted wife and collaborator of Pierre" preceded her name; stories in the press focused on her monastic years in the shed, when she and Pierre gave themselves up for their life-saving science. Curie knew that her American self was largely a figment of Meloney's imagination, but other than wielding an editorial hand when stories got "too personal," she was complicit in the end.[33]

The Tour of 1921

Meloney wove the web of publicity meticulously, each strand of Curie's life story braided to create an extraordinary American design. Her marriage provided the romance, her Spartan ways a sacrifice for nobler good. That her radium could be linked to medical research made it possible to see her as maternally benevolent. While the nation's chemical and physical societies embraced her as their own, some of the most conspicuous displays of gratitude took place at medical research facilities and medical colleges, where her discovery of radium was seen as a humanitarian act. During the course of Meloney's campaign the National Association of Medical Women made Curie an honorary member, seeing her as a saintly foremother. Heartwarming human interest stories followed. A horticulturalist, for example, named his prized strain of red roses for the woman he saw as responsible for the medical breakthrough that had saved his life.[34]

Meloney was gratified to see that within months the Radium Fund was oversubscribed. The call to raise a hundred thousand dollars had attracted more than one hundred thousand women and an impressive group of medical men who believed, as Meloney had hoped, that Curie's cause was their own. Carrie Chapman Catt, the popularly recognized pacifist and suffrage organizer, wrote a piece in *Woman Citizen* called "Helping Madame Curie to Help the World." Curie had become the expedient face of her maternalist politics. Men, too, had written checks in support of scientific innovation, medical advance, or other causes Curie ostensibly represented. Arthur Brisbane's one-hundred-dollar check to the Radium Fund did not honor a scientific mind but the best in womanhood, for he gave it in the name of Julia Ward Howe, famed writer of the "Battle Hymn of the Republic."[35]

Curie had hoped to accept the donations quietly, but Meloney wanted to satisfy the American thirst for celebrity, pull the drama from print, and parade it in the flesh. She had turned into public spectacle the taking of sealed bids from chemical manufacturers to produce the gifted radium. Now she wanted Curie to accept the radium in person, before her American admirers. While Curie agreed that a gesture of appreciation would be appropriate, she had not imagined the pomp and circumstance Meloney had in mind—in the form of a widely publicized tour, with ribbon cuttings, speeches, and ceremonies—exactly what the reclusive scientist had spent her life trying to avoid. She would lunch with the ladies of the Radium Fund Committee and visit with students at women's colleges before attending a massive reception of the American Association of University Women at Carnegie Hall. Standard Chemical and the Department of Mines also planned to host events where she would be presented with specimens of carnonite and other precious ores. She was scheduled to dedicate the mining bureau's new low-temperature laboratory, where, in the name of American ingenuity, she would push a button to set its machinery in motion.[36]

Others were quick to seize on the fascination for all things Curie. R. H. Macy and Company launched a series of books about radium to coincide with her tour, and executives of chemical companies bought space in the *New York Times* to welcome Curie to the United States. The American Museum of Natural History in New York also made preparations for the tour, opening an exhibit with photographs and specimens of the ore she had used to isolate her first batch of radium.[37] University men paid tribute to her as well, awarding her honorary degrees at the Universities of Pennsylvania, Pittsburgh, and Chicago, and at Northwestern, Columbia, and Yale. At Harvard the mood was less generous, however. William Duane, a former student at the Curie Institute, convinced university president Charles Eliot that she was unworthy of special honors. "I agree with you that Mme. Curie might possibly have some good influence while in this country on present discussions of feminism," Eliot told Meloney, but "since her husband died in 1906 Mme. Curie has done nothing of great importance."[38] Meloney defended her friend, citing studies confirming her accomplishments. When all else failed, she used maternity:

> It is hardly fair to say that Mme. Curie has [not] done anything of great importance since her husband's death in 1906. She had done nothing comparable with the discovery of radium and probably never will surpass that achievement. But she has contributed enough to science and to life to justify any praise or honor which might be bestowed upon her. She had carried on her scientific work, but the outstanding virtue of these years lies in the fact that having discovered radium and come into prominence she turned to her home as a normal mother and gave the intimate minute attention to her children which motherhood should impose.[39]

Reporters at the *New York Times* aided her case, reminding readers that Pierre Curie himself had insisted that his wife had been the more deserving recipient of their Nobel Prize. When men at Harvard, as at Princeton, refused to change their minds, Meloney simply pressed on, for men and women in many other organizations thought Curie the perfect vehicle for advancing their public profiles. The Philosophical Society, the American Chemical Society, the Mineralogical Club, and the Associations of Polish and Canadian Women organized formal gatherings in her honor. The National Institute of Social Science planned to host a formal dinner at the Waldorf, where Vice President Calvin Coolidge planned to present Curie with the institute's gold medal of honor. The grandest of displays was reserved for the White House itself: President Warren Harding was to present the radium as a gift to Curie

from American women. Politicians, industrialists, and socially important women hoped to benefit from their associations with Marie Curie.
The skepticism of a few university men could not damage her. [40]

The itinerary was "very intimidating," Curie confided. Since her arrival in France in 1891 she had rarely traveled outside the country, and
when she did it was under an assumed name to protect her privacy. Her
fear of ceremony bordered on phobia. When she won the Nobel Prize in
1903 she feigned illness so that she could accept the prize at a later date
with less fanfare.[41] Her American tour could proceed, she told Meloney,
only if she would insert days of rest in the schedule and if her daughters,
now twenty-three and sixteen, could accompany her as companions and
ready stand-ins. She asked that the voyage overseas proceed "with the
minimum of publicity," and it did until she reached the United States.
When she disembarked from the USS *Olympic* the scene was chaotic.
Student groups sang Polish, French, and American anthems, and international dignitaries and gift-bearing Camp Fire Girls all bombarded her
at once. Customs preliminaries were waived to facilitate her progress
through the crowds of admirers that had been gathering since the early
morning hours. Curie claimed to be woozy from seasickness and begged
shelter from the press, but Meloney convinced her to pose patiently for
dozens of waiting photographers. She was relieved to be whisked away
in a limousine that Louise Carnegie had arranged for her escape. When
she reached Meloney's apartment on West Twelfth Street, she asked for
seclusion. Admirers sent flowers, but she turned callers away. When she
finally emerged, a motorcar took her to the Carnegie home on Ninety-
first and Fifth, where Mrs. Cornelius Vanderbilt and socialites of the
Radium Fund Committee welcomed her to the United States.[42]

Curie graciously nodded, smiled, and shook hands, but wondered
why no woman scientist was in their midst. In the weeks that followed
she traveled from venue to venue as a great maternal figure, her science excised from her legend. A few women scientists noted the discrepancy. Untenured Columbia University instructor Christine Ladd
Franklin reminded readers of the *New York Times* that Curie was, above
all else, an exemplar of women's potential in science. Look at the accepted scientific nomenclature, she argued, for "the curie" had become
the unit of radioactivity much like the volt, the coulomb, the ohm, the
farad, the henry, and the ampere had become units of measure in other

fields.[43] But Ladd's reminders fell on deaf ears even at women's colleges, where administrators thought Curie was best used to push maternalist agendas. At Carnegie Hall, American Association of University Women members from nearly every college admitting women along the Eastern Seaboard lauded her scientific achievement as evidence of women's greater maternal influence. They, too, interpreted her work with radium as a gesture of humanitarianism rather than as evidence of scientific acumen. Bryn Mawr College president M. Carey Thomas urged her collegiate audience to pursue science if so inclined, but more important, to use science to exert influence on politics and pacifist diplomacy. American women had won suffrage months before, and Thomas hoped Curie would inspire them to form into a voting bloc to bring about disarmament—"human legislation," as she called it.[44]

Curie must have been perplexed. Never in her lifetime had she considered herself a political being of any sort, let alone a pacifist or feminist. The family of her late husband was openly radical, but she was too busy in the lab to join them as card-carrying members of the Communist Party. During World War I she and Irene volunteered to operate portable X-ray machines to treat injured soldiers on the frontlines. If this made her political, she defied someone to label her politics. Her independent choices implied a feminist sensibility, but the moment someone called her a feminist her reaction was denial.[45]

The college women she met on the tour insisted that they had much in common with her, but she didn't see it. Her college days had been Spartan and somber, but the young coeds who greeted her appeared fresh faced, well fed, and even giddy as they serenaded her with school chants. She was taken aback by their facilities—the lavish campus trees, the extensive sports equipment, the recreational rooms, and the spotless dormitories and sanitary showers. More impressive still were the clean and modern laboratories, more luxurious than hers in Paris. From her perspective the American conception of women's education was radically different from her own experience. College women seemed to be taken seriously, and yet few seemed truly impassioned by science, as she had been in her youth. At Smith College Curie hoped to meet serious-minded women, but instead she was overwhelmed by empty pomp and circumstance. The wife of Vice President Coolidge and the mayor of Northampton joined a crowd of two thousand to honor the retiring scientist. The gathering was mere prelude to the thirty-five hundred college women who greeted her at Carnegie Hall in New York City. As honored guest she sat through choral performances and yet another ceremony, this time one awarding her the Ellen Richards Memorial Prize. She was close to collapse as she tried to thank the women graciously for their medals and fleur-de-lis, but her remarks were brief and barely audible even to people ten feet away. Assuming she had said enough, her hostesses responded with rigorous applause. Curie walked away in something of a daze; by now her cataracts were so bad, she could barely see the emotion she had stirred up before her.[46]

With diminishing energy, she plodded on with her scheduled engagements. Men stopped production at radium plants so that she could consecrate the grounds, researchers at hospitals of radium therapy bowed in her presence, and society women opened their homes and pocketbooks on her behalf. The most prestigious of scientific men conferred honors no woman had won before her from the American Philosophical Society, the College of Physicians, and the American Chemical Society. The American Museum of Natural History made her a fellow, a status never before achieved by a woman. At Wellesley, hers was the first honorary degree in the history of the college. Administrators at the University of Pennsylvania announced that for the first time in school history they would hold a special ceremony to bestow the degree.[47]

The ceremony at the White House was perhaps the most difficult for her, since it had come to signify more than Curie wanted to represent. President Harding handed her a key that opened a box in which her radium would be placed, though for now it remained at the Bureau of Standards. The absence of the very radium for which she came was of no consequence to the cabinet officers, military men, Supreme Court justices, foreign diplomats, and elite scientists who convened before her. Representatives of women's societies welcomed her to the United States as its new "adopted daughter," and Harding declared the radium a symbol of the "convergence of intellectual and social sympathies" between Curie and the American people. American women, he said, offered their gift of radium much as they would shelter to the homeless or food to the hungry: "It has been your fortune, Mme. Curie, to accomplish an immortal work for humanity. . . . We lay at your feet the testimony of that love which all the generations of men have been wont to bestow upon the noble woman, the unselfish wife, the devoted mother. . . . The zeal, ambition and unswerving purpose of a lofty career could not bar you from splendidly doing all the plain but worthy tasks which fall to every woman's lot."[48]

Curie said nothing, but her companions sensed her growing anxiety; she had become everything to everyone, an icon too big and distorted to rein in. All along she had asked Meloney to tone down the publicity, particularly this fictitious association with cancer research. Meloney made sure newspapers retracted statements proclaiming that radium cured all cancers, and at Vassar College Curie timidly reminded her audience that her discovery of radium was the work "of pure science . . . done for itself" rather than with "direct usefulness" in mind. But her maternal altruism had been established. Exhausted by the misplaced attention, she increasingly let Eve and Irene accept honors on her behalf. On a train to yet another destination, she insisted that stepping foot in the public car was more than she could handle: "I cannot go in and be stared at like a wild animal," she whispered. Before reaching the Western states, she was asking to go home.[49]

Publicists found it difficult to conceal that the admiration she inspired wasn't necessarily returned. By the time the tour reached its third week, her irritation with American hospitality had become a topic for print. "Mme. Curie is 'completely tired out,'" the New York Times reported. "Questions and the 'small talk' of American women and men also have fagged her brain." The following day a journalist reported that the customary handshaking was "beginning to bore the French scientist"; she wore a sling to avoid the grips of admirers. The American press excused her behavior, supposing her aloofness to be an unfortunate symptom of radium illness, the after effects of decades of humanitarian devotion to science. "Mme. Curie is somewhat anemic, as nearly all persons of confined, studious pursuits are," the Times reported. "She had been confined most of her life to work in the laboratory. She is a woman of 53 years. With a delicate physique and unaccustomed to outdoor life, she has been attempting to put through a strenuous program in this country and it has tired her."[50]

Americans forgave, but didn't necessarily know what to make of the woman, for the press coverage on her was also a flurry of contradictions. Observers noted her tireless energy, as well as her unceasing fatigue; her confident command of English, but also her broken, inaudible whispers. Did Americans interest her or bore her? Did her "severely plain" dress suggest that she was masculine, too stuck in her head to be concerned with fashion; or feminine, too saintly to succumb to vanity? Did her "angular" face, her penetrating, deeply set gray blue eyes, and her protruding forehead betray masculine intellect or maternal warmth?[51] Observers couldn't make up their minds. In the summer of 1921, Marie Curie stood squarely at the center of conflicting thoughts about women and science.

As Curie packed her bags for France it was clear that Meloney's tour of mutual admiration had been, in the end, a calculated transaction. Curie received one hundred thousand dollars of radium, twenty-two thousand dollars of mesothorium and other precious metals, two thousand dollars from the Ellen Swallow Prize, sixty-nine hundred dollars in miscellaneous fees, fifty-six thousand dollars of leftover donations to be set aside in a fund, and a fifty-thousand-dollar advance for her impending autobiography. In return, she gave American women and chemical producers the license to manipulate her image to promote maternalist feminism and American innovation. Meloney had not placed women scientists into her plan. Aside from brief words by medical researchers Florence Sabin and Alice Hamilton at Carnegie Hall, no women scientists had been featured in her events. Curie had not spoken at primary or secondary schools, where she might have inspired impressionable young girls; aside from her visit to science students at Hunter College, none of her activities seemed to promote women scientists specifically. This was of little concern to Curie herself. Except on Irene's behalf she had never made special accommodations for women at the Curie Institute and in fact refused the young nuclear physicist Lise Meitner when she had applied for work there. Although she was one of the only women who graduated from the Sorbonne, Curie never tried to extend that distinction to others, nor did she openly challenge the sexism that plagued her own career both before and after Pierre's death. In the American press she never disclosed her keys to success, nor did she concede her disappointing defeats. American women inferred through her silence that they had to outperform men to succeed.[52]

In this we see two sides of the image-making coin: Meloney deliberately had shaped Curie into a maternal martyr who had used science for womanly ends; but this portrait also evoked an image of a superwoman, too smart, too dedicated, too focused, and too talented to be emulated by ordinary women. Trying to shape Curie in appealing ways, Meloney unwittingly kept alive popular stereotypes about women and science. To succeed in men's fields, women couldn't be themselves; they had to perform better than men, much as Curie had. Institutional sexism remained unchallenged, as women told themselves simply to work harder or publish more. Meloney had created a schizophrenic figure: a serious scientist (a masculine type) and a sacrificing woman (a maternal type), both inherently incompatible. In the minds of male employers women would always fall short of the ideal, and women who internalized their alleged deficiencies could not move forward. If Meloney's portrait of Madame Curie was supposed to inspire women, it may have done the reverse.

Postscript

Throughout the 1920s American women kept informed about their French investment. Louise Carnegie and other former committee members organized a trust fund to ensure a steady income for Curie's pursuits. Eleanor Clay Ford, daughter-in-law of Henry Ford, provided a new car, and Mrs. Henry Moses a chauffeur to take Curie to and from her lab. Meloney's associates also arranged for the overseas delivery of X-ray tubes, galvanometers, electromagnets, voltmeters, and other specialized instruments to the Curie Institute. Their devotion was unwavering, for the French scientist allowed them to realize their professional ambitions vicariously. "It is very comforting . . . but makes me feel the responsibility ahead of me," Curie confided. As American schoolteachers continued to request pictures and autographs to place in their classrooms, it was clear that Curie as symbol had transcended anything she had imagined.[53]

Curie continued to hunt for radium, this time for the Curie Institute newly installed in Warsaw. The price of it had halved since her American tour, but it was still prohibitively expensive. In 1928 the Radium Committee raised another fifty thousand dollars to purchase another gram on Curie's behalf, and thus Meloney clamored for another American tour. Curie voiced her litany of health concerns but agreed to come under stricter conditions: no long engagements, no social occasions, no interviews, no shaking of hands. Train rides in cold weather and the Girl Scouts were out of the question! The fewer appearances the better, and they had to be related to work! "Remember," she wrote

Meloney, "I am now many years older. . . . You have already forgotten how ill I had been during my visit to [the] United States in 1921." Unquestioningly, Meloney typed up a statement of her conditions to distribute to the press.[54]

Throngs of admirers were not waiting when Curie's steam liner arrived in New York in the fall of 1929. She took a back stairwell to the lower level of the pier, where a limousine took her once again to Meloney's apartment. Now Herbert Hoover would present the radium to her, or at least a check to pay for it. Reluctantly, she visited the White House and attended an event at the National Academy of Sciences building. Her only public appearance in New York City was at the annual dinner of the American Society for the Control of Cancer. Administrators at the General Electric plant in Schenectady closed down operations for a day so that she could run experiments before dedicating the new chemistry building at St. Lawrence University nearby. At the request of Henry Ford she grudgingly traveled to Dearborn, Michigan, to attend a dinner in honor of Thomas Edison on the golden anniversary of his invention of the incandescent bulb. En route to the Midwest she complained bitterly of the cold. At the dinner she refused to autograph menus, even as Ford and the president signed them beside her.[55]

American women accepted Curie's requests and tuned her schedule to suit her, yet this final U.S. tour was hardly about them at all. Aside from an event hosted by the New York Federation of Women's Clubs honoring her as a humanitarian, Curie's appearances served to reinstate masculine science. At Edison's Golden Jubilee one prominent male inventor paid homage to another, and at General Electric the inventor of the X-ray tube served as her guide through laboratories occupied exclusively by men. Alumna Emily Eaton Hepburn donated a statue of Curie to go alongside the chemistry building she dedicated at St. Lawrence, but men flanked her at the ceremonies. Physicist George Pegram spoke: "We, with the rest of the world, honor Mme. Curie for her very life, for her steadfast devotion to science, her patriotic service, her modesty. We honor her as a wife and mother. The nobility of her life is such that our admiration for her character almost turns our attention from her scientific rank."[56] On this tour, as on the last, Curie's science—and more important, the science of American women—had been erased from view. The cult of maternity overrode substantive discussion of women's work in American science.

As she sailed back to France, tour officials took precautions "to shield her from excitement," but they were not needed in the end. Upon her arrival in Europe flares went off in the Gare Saint-Lazare for two more celebrated women passengers, silent film stars Pola Negri and Alice Terry. The diversion suited Curie, for all she ever wanted to do was return to her anti-natural path. For a time she had had to step off it to keep her science alive. Reporters in the United States sensed her focus on the science much more in 1929 than in 1921. In their hands she turned less identifiable and more eccentric. In 1930 the *New York Times* described her no longer as the saintlike humanitarian, but rather as the brooding recluse, not nearly as likeable as Albert Einstein, the scientist who had since won over Americans with his charms. The *New York Times* reported that at a meeting of the League of Nations' Committee on Intellectual Cooperation, she rudely declined all social invitations after the business meeting had adjourned. She refused to be photographed or interviewed or to tell reporters what it was that kept her so occupied. Einstein, meanwhile, was congenial. Like all brilliant men he appeared a bit scattered, forgetting to light his pipe before placing it in his mouth, but his smile was engaging. He could laugh at himself and the world, unlike the stoic Curie. Yale radiochemist Bertram Boltwood reflected on her second tour and thought the unlikable woman had "made a good clean up over here." Even so, he "felt sorry for the poor old girl, she was a distinctly pathetic figure."[57]

Meloney hoped that the tours would be mutually beneficial to Curie and American women, but Curie may have been the greater beneficiary in the end. Women's opportunities in American science did not grow as a direct result of her tours. Forty-one women earned science PhDs in 1920 and 138 in 1932, three years after Curie's second tour. Fewer than one thousand women earned science doctorates over these twelve years—an increase from the previous decade, but not the substantial change many had predicted. In Curie's field of physics women averaged less than three doctorates a year, and because men, too, were taking more degrees in these years, only 5 percent of doctorates in all science fields taken together were earned by women.[58] When Curie left New York the second time, check in hand, the fallout of Black Tuesday on Wall Street was just settling in. Her departure coincided with the fiscal unraveling that devastated many sectors of the American economy, not the least of which was the chemical industry that her discoveries and image had propped up and that American women had hoped to enter. Women scientists who had found jobs in laboratories after Curie's first American tour relinquished them to men after the second. Their professional aspirations appeared inappropriate amid pressures to contribute to shrinking family wages, and thus many of them turned to pink-collar sectors for employment.

During the Depression, walking Curie's path seemed an impossible dream. She represented women's untapped potential less and less and became more and more clearly a superhuman anomaly. While young American women might admire Curie's accomplishments, they would see her simplicity and asexual appearance and behavior as unappealing relics of a bygone Victorian era. If they were to emulate any Curie, it was not likely to be Marie or Irene but Eve, the fashionable cosmopolitan artist who wore lipstick and had sex appeal. The sociologist Lorinne Pruette confirmed that even by 1924, Marie Curie's name had no resonance for teenaged girls of the flapper age. When 347 teenagers were asked to name a heroine, real or literary, who inspired them, only three girls chose Curie, the more glamorous Cleopatra beating her out thirty times over. Respondents who desired careers chose social science before physical science and artistic careers above all (46 percent), further underscoring how impractical and unwomanly they perceived Curie's choices for their own lives. In the postsuffrage era women with professional aspirations wanted to combine marriage and career, and yet 89 percent of the women who appeared in *American Men of Science* in 1927 were unmarried, and in Curie's field of physics, the percentage was even higher. The French woman seemed, in her later years, to be—and perhaps to have been—a joyless spinster who had chosen science over sexuality, and the trade-off was not especially appealing.[59]

In the end the Curie tours were not useful for women in science. Sex-typed employment remained constant in scientific sectors, and any headway women had made in science fields as a result of World War I was afterward virtually erased, to the seeming indifference of Curie herself. American women had to wait for the next world war to make new gains—only to suffer yet another closing of opportunities in its aftermath. Fatigued by their failure to make inroads in the upper echelons of science, women assumed a nonconfrontational stance, characterized, according to Margaret Rossiter, by deliberate overqualification and personal stoicism.[60] Even the most accomplished women scientists in the

1950s and 1960s continued to insist that the key to success was to perform head and shoulders above male competition, sacrificing family, health, and sanity for research. This remained the unconscious strategy of women scientists until second-wave feminists began to see the Curie complex for what it was: a mirage that kept women from making headway in science in any substantive terms.

The Curie tours of the 1920s were an opportunity that had come and gone—and women scientists sensed it. In 1921 industrial engineer Lillian Gilbreth and astronomer Cecilia Payne-Gaposchkin imagined bright prospects in their respective fields of science. Both were products of doors opening in American universities: Gilbreth at Berkeley and Brown, and Gaposchkin at Harvard. But the expansion of science training for women in universities did not translate into an expansion of opportunities in scientific work. Gaposchkin first appeared at the Harvard observatory in a period when the old guard of amateur women "computers" was on its way out and a new breed of professional woman astronomer was emerging. In these years she got her foot in the door, but she also hit a glass ceiling at almost the same time. She discovered that in the university, women either remained on the lower rungs of the promotional hierarchy or were ghettoized in feminized fields. Gilbreth experienced no less hostility in the industrial sector. Initially an engineering consultant who publicly (though not actually) was an assistant to her husband, she found it difficult to maintain their business when he died tragically in 1924. She, like so many women, had to work within feminized niches of science where the male establishment offered little resistance. Her choices did not necessarily signify failure, for she found rewarding her work on efficiency in schoolrooms, department stores, and domestic spaces, and her impact on the lives of women and families was palpable and ultimately lasting. She and Gaposchkin reveal the mixed legacy of the 1920s, when women suffered dashed professional dreams, but found ways to succeed on their own satisfying terms.

Notes

Note on *New York Times* citations: In cases when original page numbers were not visible on scanned articles, the page numbers are those provided by ProQuest Historical Newspapers.

1. Marie Mattingly Meloney, introduction, to Marie Curie, *Pierre Curie* (New York: Macmillan, 1923), 12–16.

2. *Vanity Fair*, December 22, 1904; Emily Crawford, "The Curies at Home," *World To-Day* 6 (1904): 490.

3. "Women and Scientific Research," *Science* 32 (December 23, 1910): 919–20; "Sex and Scientific Recognition," *Scientific American* 104 (January 21, 1911): 58; Margaret Rossiter, *Women Scientists in America: Struggles and Strategies to 1940* (Baltimore: Johns Hopkins University Press, 1982), 108–9.

4. Marcel LaFollette, *Making Science Our Own: Public Images of Science, 1910–1955* (Chicago: University of Chicago Press, 1990), 79–80; Benjamin Harrow, *Eminent Chemists of Our Time* (New York: D. Van Nostrand, 1920), v, 165.

5. Meloney, introduction, 17.

6. "The Story of Radium," *New York Times* (hereafter *NYT*), May 15, 1921, 84; Robert Reid, *Marie Curie* (London: Collins, 1974), 160, 247; Naomi Pasachoff, *Marie Curie and the Science of Radioactivity* (New York: Oxford University Press, 1996), 65.

7. Meloney, introduction, 15–16.

8. Einstein came to the United States to raise money for the Hebrew University in Jerusalem. Don Arnald, "Einstein on Irrelevancies," *NYT*, May 1, 1921, 50; Daniel Kevles, *The Physicists: The History of a Scientific Community in Modern America* (Cambridge: Harvard University Press, 1987), 212.

9. Kevles, *The Physicists*, 206; "Roll of Martyrs to Science Is Increasing," *NYT*, January 11, 1925, XX5.

10. Judith Magee, "Marie Curie: A Study of Americans' Use of Marie Curie as a Devoted Wife and Mother, Saintly Scientist, and Healer of Humanity" (master's thesis, Sarah Lawrence College, 1989), 3–7 to 3–8; Reid, *Marie Curie*, 53–54.

11. Thomas C. Jeffries, "The Story of Radium in America," *Current History Magazine* 14 (June 1921): 2; William J. Mayo, "Do You Fear Cancer?" *Delineator* 98 (April 1921): 35.

12. G. Frank, "A Jeanne D'Arc of the Laboratory," *Century* 102 (May 1921): 160; "Madame Curie: The Most Famous Woman in the World," *Current Opinion* 70 (June 1921): 760–62; Marie Mattingly Meloney, "The Greatest Woman in the World," *Delineator* 98 (April 1921): 15–17; Magee, "Marie Curie," 2–23, 3–11. See also Nell Ray Clarke, "Radium—the Metal of Mystery: How Madame Curie Found It," *American Review of Reviews* 63 (June 1921): 606–10; "Mme. Curie, Widow of Prof. Pierre Curie, and with Him Co-Discoverer of Radium with her Two Daughters," *American Review of Reviews* 63 (June 1921), 562; "Curing Cancer with Radium," *Current History Magazine* 13 (October 1920): 34–36; "Madame Curie's Own Account of the Radioactive Elements," *Current Opinion* 70 (May 1921): 656–58; "She Discovered Radium, But Hasn't a Gram of It," *Literary Digest* 69 (April 2, 1921): 36–40; "Her Discoveries Made Others Rich," *Mentor* 9 (May 1921): 36; "The Discoverer of Radium," *Outlook*, May 25, 1921, 145; "Madame Curie's Visit to America," *Science*, April 8, 1921, 327–28.

13. Meloney, introduction, 23; "Madame Curie's Genius," *NYT*, May 1, 1921, 103.

14. Marie Curie, "Autobiographical Notes," in *Pierre Curie*, 170–72; "Madame Curie's Genius"; Harrow, *Eminent Chemists*, 156; Bernard Jaffe, *Crucibles: The Lives and Achievements of the Great Chemists* (New York: Simon and Schuster, 1930), 246; Eleanor Doorly, *The Radium Woman: A Youth Edition of the Life of Madame Curie* (Hammondsworth, U.K.: Puffin Story Books, 1953), 78–85 (original edition, Puffin Story Books, 1939); Reid, *Marie Curie*, 48–49; Rosalynd Pflaum, *Grand Obsession: Madame Curie and Her World* (New York: Doubleday, 1989), 18–30.

15. For analysis of the Curies' "anti-natural path," see Helena M. Pycior, "Marie Curie's 'Anti-natural Path': Time Only for Science and Family," in *Uneasy Careers and Intimate Lives: Women in Science, 1789–1979*, ed. Pnina G. Abir-Am and Dorinda Outram (New Brunswick: Rutgers University Press, 1987), 191–214. "Love Story of Famous Mme. Curie

Recalls Romance of School Days," *NYT*, May 27, 1923, XX5; Curie, *Pierre Curie*, 74–77, 86–90; Jaffe, *Crucibles*, 247–48; Doorly, *Radium Woman*, 88; Reid, *Marie Curie*, 56–65, 68–69.

16. Curie, *Pierre Curie*, 80–82; Jaffe, *Crucibles*, 248; Doorly, *Radium Woman*, 95, 102–3; Reid, *Marie Curie*, 70, 73; Pflaum, *Grand Obsession*, 56.

17. Curie, *Pierre Curie*, 96–101; "Autobiographical Notes," 180–89.

18. Harold Callender Paris, "Mme. Curie Works on with Her Pupils," *NYT*, December 23, 1928, 11, 15; James Kendall, *Young Chemists and Great Discoveries* (New York: D. Appleton-Century, 1940), 214; Curie, *Pierre Curie*, 101–4; Jaffe, *Crucibles*, 251–52; Doorly, *Radium Woman*, 116; Reid, *Marie Curie*, 85, 96–97, 106; Pflaum, *Grand Obsession*, 66–69.

19. Curie, "Autobiographical Notes," 178–79; Pflaum, *Grand Obsession*, 62.

20. Harrow, *Eminent Chemists*, 174–75; Doorly, *Radium Woman*, 127; Reid, *Marie Curie*, 84–85, 92.

21. Reid, *Marie Curie*, 92; Doorly, *Radium Woman*, 127.

22. Reid, *Marie Curie*, 127–29; Pflaum, *Grand Obsession*, 101; Curie, "Autobiographical Notes," 179; Pasachoff, *Marie Curie*, 32–33.

23. Curie, "Autobiographical Notes," 176; Reid, *Marie Curie*, 131.

24. Curie, "Autobiographical Notes," 193–94; Reid, *Marie Curie*, 170; Pflaum, *Grand Obsession*, 78, 82, 150, 270; Pasachoff, *Marie Curie*, 82.

25. Curie, *Pierre Curie*, 113; Reid, *Marie Curie*, 74.

26. Pasachoff, *Marie Curie*, 54–55; Reid, *Marie Curie*, 132–33; Harrow, *Eminent Chemists*, 161.

27. Ayrton, quoted in Reid, *Marie Curie*, 219, 82; Helena M. Pycior, "Pierre Curie and 'His Eminent Collaborator Mme Curie,': Complementary Partners," in *Creative Couples in the Sciences*, ed. Helena M. Pycior, Nancy G. Slack, and Pnina G. Abir-Am (New Brunswick, NJ: Rutgers University Press, 1996), 47; Curie, "Autobiographical Notes," 96–97.

28. Harrow, *Eminent Chemists*, 169; Curie, "Autobiographical Notes," 192.

29. Meloney, introduction,13–14; Curie, *Pierre Curie*, 141; Harrow, *Eminent Chemists*, 174; Jaffe, *Crucibles*, 259; Kendall, *Young Chemists and Great Discoveries*, 218; Doorly, *Radium Woman*, 148; Reid, *Marie Curie*, 155–56; Pflaum, *Grand Obsession*, 141.

30. "L'Académie des Sciences examine aujourd'hui la candidature de Mme Curie," *L'Excelsior*, January 9, 1911, cover; Reid, *Marie Curie*, 179–205; Pflaum, *Grand Obsession*, 156–74; Pasachoff, *Marie Curie*, 67–73.

31. Reid, *Marie Curie*, 211–14; Pasachoff, *Marie Curie*, 77; excerpts from Curie's 1911 acceptance speech are from Eduard Farber, *Nobel Prize Winners in Chemistry, 1901–1950* (New York: Henry Schuman, 1953), 47.

32. Magee, "Madame Curie," 3–8; Reid, *Marie Curie*, 256.

33. Marie Curie to Missy Meloney, November 7, 1920; September 15, 1921, Box 1, William B. Meloney-Marie Curie Special Manuscript Collection, Columbia University Libraries, New York, New York (hereafter MCC).

34. "Radium Not a Cure for Every Cancer," *NYT*, May 13, 1921, 22; "Medical Women Honor Mme. Curie," *NYT*, June 12, 1921, 14.

35. Carrie Chapman Catt, "Helping Madame Curie to Help the World," *Woman Citizen* 5 (March 12, 1921): 1062; Arthur Brisbane to Gentlemen, March 14, 1921, Box 1, MCC.

36. Meloney, introduction, 20–21; Marie Curie, "Impressions of America," Box 3, MCC; "Mme. Curie at Dedication," *NYT*, May 22, 1921, 2.

37. Ad of R. H. Macy, *NYT*, May 7, 1921, 4; Ad of Radium Luminous Material Corporation and Radio Chemical Corporation, *NYT*, May 12, 1921, 14; "Show Curies' Work by Radium Exhibit," *NYT*, May 22, 1921, 32.

38. The first woman to receive an honorary degree from Harvard was Helen Keller in 1955. "Mme. Curie in Boston," *NYT*, June 19, 1921, 25; Charles W. Eliot to Marie Mattingly Meloney, December 18, 1920, Box 2, MCC.

39. Marie Mattingly Meloney to Charles W. Eliot, December 24, 1920, Box 2, MCC.

40. "Madame Curie's Genius"; "Mme. Curie Plans to End All Cancers," *NYT*, May 12, 1921, 1, 3.

41. Curie, *Pierre Curie*, 125; "Autobiographical Notes," 229; Pflaum, *Grand Obsession*, 115.

42. Marie Curie to Meloney, January 31, 1921, Box 1, MCC; "Mme. Curie Feted at Carnegie Home," *New York Herald*, European edition, May 13, 1921; "Mme. Curie Plans to End All Cancers"; "Mme. Curie Here Today," *NYT*, May 11, 1921, 8; "Mme. Curie Lands, Cheered by Crowd," *NYT*, May 12, 1921, 1; Marie Curie, "Impressions of America."

43. Christine Ladd Franklin, "Mme. Curie and the Curie," *NYT*, June 4, 1921, 9.

44. "Says Women Can and Must Stop War," *NYT*, May 19, 1921, 11.

45. Irene Curie to Missy Meloney, November 11, 1934, Box 2, MCC.

46. Curie, "Impressions of America"; "Smith College Gives Degree to Mme. Curie," *NYT*, May 14, 1921, 9; "Says Women Can and Must Stop War."

47. "Mme. Curie Called Greatest Scientist," *NYT*, May 18, 1921, 34; "Exception for Mme. Curie," *NYT*, May 18, 1921, 14; "Award Mme Curie a Wellesley Degree," *NYT*, June 21, 1921, 7.

48. Curie, "Impressions of America"; "Radium Presented to Madame Curie," *NYT*, May 21, 1921, 12; "President Harding's Speech to Madame Curie," *New York Herald*, European edition, May 21, 1921, 1.

49. "Radium Not a Cure for Every Cancer"; Magee, "Madame Curie," 3–12; "Honors for Mme. Curie Received by Daughter," *NYT*, May 24, 1921, 17; "Mme. Curie Again Honored," *NYT*, May 27, 1921, 16; "Mme. Curie to Start West," *NYT*, May 31, 1921, 6; Pflaum, *Grand Obsession*, 206–207; Pasachoff, *Marie Curie*, 88; Reid, *Marie Curie*, 262, 266.

50. "Mme. Curie's Brain Fagged by 'Small Talk' of Americans," *NYT*, May 28, 1921, 1; "Memorial Hospital Greets Mme. Curie," *NYT*, May 29, 1921, 16; Reid, *Marie Curie*, 169; Pflaum, *Grand Obsession*, 152.

51. Mme. Curie Sails to Receive Radium Gift," *NYT*, May 5, 1921, 14; "Madame Curie's Genius"; "Mme. Curie to End All Cancers."

52. "Scientists' Medal Given to Mme. Curie," *NYT*, May 20, 1921, 3; "Plan Life Income Now for Mme. Curie," *NYT*, July 30, 1921, 7; Magee, "Madame Curie," Appendix, Table 4, "Curie's Awards in 1921"; Rossiter, *Women Scientists in America*, 100, 124.

53. "Plan Life Income Now for Mme. Curie"; Louise W. Carnegie to Mrs. William Brown Meloney, January 23, 1930; February 6, 1930; Marie Curie to Meloney, March 17, 1922; May 20, 1922; January 27, 1930; February 18, 1930, Box 1; Meloney to Mrs. S. L. Quimby, March 27, 1922; Meloney to Marie Curie, January 15, 1930, Box 2, MCC.

54. Marie Curie to Meloney, August 21, 1928; July 28, 1929; n.d. 6, 1929, Box 1, MCC.

55. "Mme. Curie Arrives 'Happy to Be Back,'" *NYT*, October 16, 1929, 30; "Mme. Curie Is Guest of Friends in Country," *NYT*, October 17, 1929, 9; "Cancer Group Here to Honor Mme. Curie," *NYT*, October 20, 1929, N1; "Mme. Curie Examines Schenectady Plant," *NYT*, October 24, 1929, 3; "Mme. Curie Serenaded," *NYT*, October 26, 1929, 21; "Mme. Curie Speaks at St. Lawrence," *NYT*, October 27, 1929, 24; "Mme. Curie at White House," *NYT*, October 30, 1929, 21; "Madame Curie Has a Cold," *NYT*, October 28, 1929, 22; "Honor Madame Curie Tonight," *NYT*, October 31, 1929, 17; "Mme. Curie Receives $50,000 Radium Gift; Hoover Presents It," *NYT*, October 31, 1929, 1; "Mme. Curie's Aversion to Giving Autograph Extends Even to Endorsing Charity Checks," *NYT*, February 15, 1931, 54.

56. "Mme. Curie to Get Medal," *NYT*, October 21, 1929, 16; "Clubwomen Give Mme. Curie a Medal," *NYT*, November 6, 1929, 27; "Mme. Curie Speaks at St. Lawrence."

57. "Movie Stars Efface Mme. Curie's Return," *NYT*, November 16, 1929, 5; "Einstein Evolving Yet Another Theory," *NYT*, July 27, 1930, 1; Boltwood, quoted in Reid, *Marie Curie*, 268.

58. Kevles, *The Physicists*, 204–6.

59. Lorine Pruette, *Women and Leisure* (New York, 1924). More on Pruette's findings in William O'Neill, *Everyone Was Brave: A History of Feminism in America* (New York: Quadrangle, 1971), 322; Luella Cole Pressey, "The Women Whose Names Appear in 'American Men of Science,' for 1927," *School and Society* 29 (January 19, 1929): 96–100; Kevles, *The Physicists*, 205.

60. Margaret Rossiter, *Women Scientists in America*, 100, 130.

2 Making Science Domestic and Domesticity Scientific: The Ambiguous Life and Ambidextrous Work of Lillian Gilbreth

> It was Mother the psychologist and Dad the motion study man and general contractor, who decided to look into the new field of the psychology of management, and the old field of psychologically managing a household of children. They believed that what would work in the home would work in the factory, and what would work in the factory would work in the home.
>
> —Frank Gilbreth Jr. and Ernestine Gilbreth Carey,
> *Cheaper by the Dozen*, 1948

> The family, what is it but an experiment. The Quest of the One Best Way!
>
> —Lillian Gilbreth, *The Quest of the One Best Way*, 1925

> Here I may be a Scientist
> Who measures as she makes
> Here I may be an Artist
> Creating as she bakes
> Here busy heart and brain and hand
> May feel and think and do
> A kitchen is a happy place
> To make a dream come true.
>
> —Promotional pamphlet for *Lillian
> Gilbreth's Kitchen Practical*, 1931[1]

ON JUNE 17, 1924, IN MONTCLAIR, NEW JERSEY, A MOTHER OF TWELVE SAT at the table with her children for their most urgent "Family Council." The topic of discussion was not allowance or curfews, but how to proceed now that the patriarch of the house, industrial engineer Frank Gilbreth, was no longer present and in charge. Just a few days ago he had died of a heart attack in a phone booth at the local train station. It was the same day that his second-eldest daughter was graduating from high school, and he was rushing around before the ceremony to get visas processed for an upcoming trip across the Atlantic, to attend the World Power Conference in England and the International Management Congress in Prague. He and his wife planned to represent the American Management Association at these events, but he had suddenly felt worried about the passports. So he stopped to call Lillian and asked her to check his desk drawer. She did, and when she picked the receiver back up, he was no longer on the line. The train must have arrived, she thought, but bystanders confirmed that Frank had collapsed and died instantly on the platform.[2]

Six years earlier, when he had fallen ill while consulting for the U.S. military, doctors had warned that Frank's heart wouldn't take his breakneck pace. At Fort Sill he organized the rehabilitation of disabled soldiers and he taught active soldiers how to clean and assemble Lewis and Browning machine guns with a minimum of physical motion and muscle fatigue. But there he was struck with rheumatism, then uremic poisoning, followed by a bout of pneumonia that kept him bedridden for months. He recovered, but his heart had been compromised. Lillian made sure he dieted and carried a heart stimulant at all times, yet he seemed to gain more and more weight as Lillian turned more waiflike. When he collapsed in 1924, he weighed nearly 230 pounds.[3]

Fearing the worst hadn't prepared Lillian for the tasks of raising the children and paying the bills alone. Her youngest, Jane, was only a baby, and the next-oldest boys were toddlers; except for Anne, all the children still lived at home. They had been prepared for college, but the older girls now contemplated taking time off to make ends meet. The Mollers, Lillian's parents, wondered why their daughter had chosen a life of less privilege and more chaos than that to which she had been accustomed. They offered to take the children, but Lillian refused. She was not a victim; the life she had chosen was not second-rate but the path less traveled, and she felt privileged to have traveled it with a man who hadn't insulted her by expecting less. "I have over twenty perfect years to remember," she told her mother. "I have had the best. . . . He will wait for me." As daunting as the future seemed, she hadn't regretted the choices she had made.[4]

Lillian planned Frank's service without flowers or music and donated his brain to Harvard Medical School, as he had wanted. The children agreed that she should leave immediately for London to appear in their father's place at the World Power Conference. They had their standing orders, they assured her: each older child knew to watch over a younger one, and their system of charts and files would ensure that the household chores got done. Anne and Ernestine would round up the children and take them to the family's home in Nantucket for the summer. The next eldest, Martha, would take over the family finances while Mom took over Dad's speaking engagements. But no one had planned for what happened next: nearly all the children contracted chicken pox within days, and the youngest four got the measles; the baby was delirious. The family was in unprecedented crisis, and yet Lillian had faith in the unflappability of the domestic system Frank had devised. In the end she presided over his sessions in Europe and gave his papers on the Gilbreth brand of "motion study."[5]

This part of Lillian Gilbreth's story is familiar to Americans who read the book or saw the Hollywood film *Cheaper by the Dozen* after World War II. According to the story written by Ernestine and Frank Jr., Dad's death wasn't a crossroads in Mom's life, but the end of her existence in the shadow of a more charismatic spouse. The tale solidified her status in the popular consciousness as the proverbial "woman who lived in a shoe," but scholars since have also recognized something subversive underneath. Indeed the Gilbreth children played out the fantasy of traditional patriarchal authority in their recollections, but this authority was subtly undermined by the quieter, more effective control wielded by a mother who, while not the protagonist of their account, was implicitly a partner of equal standing.[6]

Readers knew her anxieties as a widow, but not the extent of her dilemmas as the living half of a business and scientific partnership. Together, Frank and Lillian had run an engineering consulting firm; given scores of professional papers; published eight books and hundreds of articles; run a summer school on industrial management; and maintained a household of servants, schedules, and a dozen children. To the outside world Lillian appeared the helpmeet to a more qualified scientific husband. But in the Gilbreth household everyone understood that Mom was an industrial consultant, a credentialed psychologist, an originator of ideas, a writer of texts, a runner of motion-study experiments, a supervisor of homework, and a soother of scrapes and hurt feelings all rolled into one.

Lillian pronounced her child-raising years with Frank the most significant period of her life, and yet her leaving for Europe in 1924 marked the beginning of a substantially longer and gratifying chapter. Her children didn't predict it: "Before her marriage, all Mother's decisions were made by her parents. After her marriage, the decisions were made by

Dad. It was Dad who suggested having a dozen children, and that both of them become efficiency experts. If his interests had been in basket weaving or phrenology, she would have followed him just as readily."[7] But with Frank gone and children to feed, Lillian followed her own professional compass. For fifty years after Frank's death she continued to be a mother, but not always a hands-on one. She reinvented herself as a civil servant; a popular advice-giver; a professional consultant and educator; a domestic icon; and, as all of these, a new brand of scientific practitioner—one who challenged established boundaries between science and domesticity, male and female spheres. Hers was a path with no blueprints, and thus it is difficult to classify the kinds of science and feminism she espoused in her career, if we can call them science or feminism at all. Rather than focus on classification, perhaps we will understand more if we study the ways in which Lillian negotiated through stereotypical labels of her time.

The daughter of a wealthy San Francisco businessman, Lillie Evelyn Moller was shy and intensely bookish, a cause for her parents' concern, since they feared she would limit her prospects for a good marriage. They balked at the idea of college, for at the turn of the century higher education was for the career minded, and a career was only for women who hadn't the privilege of being cared for. Cultivated ladies could be avid readers, musicians, and seasoned travelers, but they did not pursue formal studies with a profession other than teaching in mind—and teaching was to be relinquished upon marriage. The Moller men were providers who deferred to the Moller women on all things domestic, rarely feeding or changing the babies of the household for fear of the emasculating effects.

Lillie looked the part of the well-groomed socialite, but she identified more with such contemporaries as Jane Addams and Marie Curie: she was a straight-A student who sought a purpose outside the home. Reluctantly, her parents let her attend Berkeley to study English, but she also took philosophy and psychology and justified the indulgence as necessary equipment for her teaching certification. Her parents were shocked when she insisted on renaming herself the more sophisticated "Lillian" and on moving to New York to enroll as a graduate student in philosophy and comparative literature at Columbia. When a literary scholar refused to accept female students, she returned to Berkeley to focus on Elizabethan literature. With a master's degree in hand in 1902, she prepared for her doctorate, this time in psychology. But before beginning a life in academe, she traveled once more to Europe, her chaperone for the fashionable trip the cousin of a man whose red Winton Six Coaching Car immediately caught her eye. Its driver was the consummate bachelor Frank Bunker Gilbreth.[8]

Frank and Lillian seemed to have almost nothing in common when they met in 1903. She was a tall, slim society girl breaking away from her cloistered life; Frank was a thickset man ten years her senior and rough around the edges. She was painfully shy and soft-spoken; he boasted ad nauseum of the name he was making for himself in the gritty world of construction. As Lillian told it years later, his story was truly one of self-making. Just shy of seventeen, he had been poised to enter MIT, but studying didn't appeal to his restless spirit. He became a bricklayer's apprentice and took mechanical engineering courses at night and within two years was a foreman, two years later a superintendent. As he mastered the bricklaying trade he developed the quickest techniques for handling corners and built rigging to minimize needless movement and lifting. Traditional methods required as many as eighteen compartmentalized movements to lay a brick; he had whittled them down to five, allowing a man to double his hourly yield. After patenting his adjustable scaffold, he invented concrete mixers, conveyors, and other apparatuses that allowed him to build houses, mills, canals, skyscrapers, and the basic infrastructure of whole towns in record speed. He started his own construction company in 1895 and went back to MIT to orchestrate the building of its electrical laboratory in a stunning eleven weeks.[9]

Frank Gilbreth and Company had grown to be one of the largest construction firms in the country by the time Frank met Lillian in 1903. They courted for a year and married in San Francisco, honeymooning at the St. Louis World's Fair. From the beginning Lillian could see Frank's penchant for efficiency: her engagement ring was engraved with the date of their engagement before he had even popped the question. He had already started apprenticing his wife on the train ride to St. Louis; passengers watched as he pulled out pen and paper and gave her a lesson in masonry. If it seemed strange that a man reared in late nineteenth-century New England considered sharing his professional endeavors with his wife, one has to recall that Frank had not grown up with typical women. His sisters were his academic superiors; one had become a musician and the other a botanist, both world class. Frank had watched them cultivate their talents outside the home and swore that should he ever marry, his wife would do the same. He asked Lillian to show him a list of qualifications she brought to their new partnership—a "job analysis" that she later compared to the "surveying and outfitting" performed by an engineer as he first meets a client. Like any engineering contract, Frank's marriage required an assessment of assets so that the partners could "meet the world as a firm."[10]

"The One Best Marriage was to be secured through The One Best Way": The Efficient Partnership of Lillian and Frank Gilbreth, 1904–1912

At a ceremony recognizing Lillian as the first female honorary member of the Society of Industrial Engineers, a beaming Frank joked that his success had come from the sweat of his frau. There was truth to the wisecrack. He swore that had he known how fruitful their partnership would be, he would have married her immediately rather than wasting idle months in courtship. Their regular conceiving of children thereafter was, no doubt, the efficient man's way of making up for lost time. He contended that children could literally be "cheaper by the dozen," and Lillian agreed to test his theory, conceiving a child every year and a half for the next eighteen years. When all was said and done, she had been efficient indeed, birthing twelve children in thirteen pregnancies. Six were boys and six were girls, just as Frank envisioned. She nursed each baby, and other than her youngest, bore them at home in ritualized fashion. "She'd supervise the household right up until each baby started coming," the older kids recalled. "She had prepared all the menus in advance, and the house ran smoothly by itself during the one day devoted to delivery."[12]

The Gilbreth union was a sort of companionate marriage, a partnership in business and parenting—in science as much as domesticity—but not without some initial adjustments. In his thirty-six years of bachelorhood Frank had become fixed in his ways and wasn't eager to relinquish his central place in the lives of his mother and Aunt Kit, who waited on him "by inches." Lillian marveled that they never asked Frank to raise a finger, that after a day's work he always returned to "a smiling welcome, a bountifully spread table, keen admiration." His "duplicate mothers" sought a suitable apartment on the Upper West Side of Manhattan for all four of them to move into after the honeymoon, and they furnished it fully, forcing Lillian to put her items in storage. When Frank left for long business trips, Lillian was compelled to fend for herself and often was reduced to tears.[13]

As an ex–boardinghouse matron who cooked Frank's favorite meals and darned his socks, Martha Gilbreth would always be her son's beloved domestic whiz. But Lillian learned to make the situation work for her. Once Lillian bore Anne in 1905 and Mary in 1906, Martha's insistence on running the house allowed Lillian to focus solely on the children and engineering. She had a lot to learn on both fronts, but Frank could see that she had a unique perspective to offer. One of her first days apprenticing at a building site proved enlightening: the contractor watched the stone and Frank watched the masons' motions; Lillian stood watching the mason, trying to decide if he truly liked the work. Frank could see that in her studious reserve, she had become observant of people and instinctively aware of what made them tick.[14]

Frank appreciated Lillian's attention to the "human element," but it was hard to make a case for its incorporation into the field of scientific management, to which he aspired. The undisputed originator of the discipline, Frederick Winslow Taylor, had earned acclaim in the 1880s for time studies that set standards for per hour production at industrial plants. When he could replace human workers with cheaper, faster machines, he did. When the human element could not be removed, he turned people into smaller cogs of a bigger industrial machine. Once a worker had been a skilled artisan, a master of his domain; but Taylor made him exploited and expendable, with no power to defend his turf—all apparently in the name of science. Taylor's streamlined processes worked because human contingency had been removed from them.[15]

Frank defended Taylor in a primer on scientific management in 1912; Lillian helped to write it but was never awestruck by Taylor's ideas.[16] Her subjective observation of human needs had no place in his disembodied processes, nor did her interest in the "unscientific" fields of management and education. With Frank's blessing she convinced Berkeley administrators to let her resume doctoral research in educational psychology, namely on theories of habit formation in industrial workers. Ergonometric chairs and workbenches, adequate lighting, incentives, and regular breaks were good for efficiency, she insisted, but they were also of benefit to the physical and emotional well-being of the worker. Lillian studied in California until she was ready to deliver her third baby girl. In 1909 the Gilbreths left their Manhattan apartment for Plainfield, New Jersey, where the expanding family had more room. Lillian left her daughters in the care of Martha and a German nursemaid while she conducted field research at the sites of Gilbreth contracts.[17]

Frank approved, but didn't dissuade his wife from putting aside her thesis in order to help him establish himself in the efficiency movement. She indexed a new edition of *Field System* and soon was doing the lion's share of the preparation of manuscripts and papers for Frank's academic and professional meetings. Initially Taylor thought the Gilbreth writings worthy contributions to his movement, borrowing heavily from Frank's bricklaying studies to write *Principles of Scientific Management* in 1911. But he also found the Gilbreths more rebellious than his other disciples, since their means for achieving efficiency revolved less around the study of *time* than around that of *motion*. If one could cut down the movements required to perform a task, they claimed, the result was not just quickness but less worker fatigue. The key was breaking down all tasks into basic elements called "therbligs" (nearly Gilbreth spelled backward). Whether laying a brick or typing a memo, all workers engaged in some sequence of searching, finding, selecting, grasping, positioning, assembling, using, disassembling, inspecting, transporting, loading, prepositioning, releasing, waiting, resting, or planning. By reducing the number of therbligs in a process, they reduced the motions and necessarily the time and physical and material resources required to complete a task.[18]

Thus while Taylor collected data with a stopwatch, the Gilbreths relied on visual images in the form of micromotion films and cyclegraphs. They recorded workers' movements and placed a special clock, a microchronometer, in view to indicate the expenditure of time in fractions of seconds. They marked off workspaces into four-inch squares or photographed this cross sectioning onto film so that workers' movements could be measured spatially when projected onto screens and studied under magnifying glass. Data from their films was recorded onto Simultaneous Motion Cycle (or "simo") charts that revealed when therbligs were needlessly duplicated, dispensable, or performed simultaneously by other body parts. Drawing on recent innovations in chronophotography, the Gilbreths strapped lights onto the limbs of workers and captured on time-exposed photographs the paths of light created as workers performed their tasks. The shorter and more fluid the lines, the more efficient the movements. Negatives viewed through a stereopticon revealed motions in three dimensions; and because they set up their cyclegraph with an interrupter that made the lights appear to flash at a known rate per second, they could count white dots on their photographs, measure distances between them, and determine workers' speed and relative acceleration with greater precision than with a stopwatch alone.[19]

The Gilbreths boasted that they had found "the One Best Way"— the least taxing method to move the fewest body parts through the smallest space quickly. Critics called their chronocyclegraphs gimmick, not science, but Frank and Lillian insisted that their methods achieved precision and were thus more scientific than Taylor's. Moreover, they were humane. Their emphasis on diminishing *fatigue* put them in the business of conservation, not exploitation. Motion study, as they perfected it, did not strip workers of autonomy because it brought skill and pride to their work; in fact since workers participated in the micromotion films, they became integral members of the Gilbreths' investigative team. Ideally, workers also benefited from the efficiency they created, as profits got passed down to them. Amid unionists' mounting criticism of scientific management's dehumanizing effects, the Gilbreths claimed to achieve better science and humane working conditions all at once.[20]

The Gilbreths described facets of the "One Best Way" in *Concrete System* (1908), *Bricklaying System* (1909), and *Motion Study* (1911), books published under Frank's name only. Lillian understood that, in a field defined and dominated by men, her inclusion would undermine their ideas; yet the reality was that she worked so closely with Frank on publications that even he thought it impossible to tease out his contributions from hers. They stayed up late into the night laying out drawings and galley sheets, often handing off babies as they worked. "I never had so much fun as I have planning these things with you," he told her. "You are always such an inspiration and always go me one better on all my ideas." To others she claimed only to provide snacks and pillows to make Frank comfortable as he wrote, but quietly she drafted whole books in his absence. *Fatigue Study* was hers, as were other works that stressed the conservation of human energy over maximized production. In the end, the things that were most important to her won the Gilbreths the reputation as the "good exception" in the eyes of organized labor.[21]

On occasions when Lillian accompanied Frank on business, she was typically pregnant or with a baby in her arms and had left the older children at home with Martha. But for most of the tens of thousands of miles Frank traveled each year, Lillian stayed behind to run the business and household alone; she accepted that her work lay "behind the scenes," or at least that it should seem that way. Her letters to Frank in these times serve as remarkable windows into her multifaceted consciousness as mother and wife, professional and engineer; she devotedly discussed the problems of clients as well as the birthdays, first steps, and milestones Frank missed at home. His letters were affectionate, but also laundry lists of orders: "I'll begin by planning your work so that you can do it with less fatigue and in less time," he prefaced from Germany, and then itemized detailed instructions for files and correspondence. Item 16: "Take a *real* vacation before doing anything. You need it badly." The reminder was typical but farcical and never at the top of the list.[22]

Lillian may have been overly accommodating, but she felt joy to be part of Frank's vision and grateful also to have purposeful work. Frank increased her responsibilities in the consultancy, although medical men had long alleged that intellectual pursuits posed a threat to women's reproductive health.[23] Clearly, Lillian would have been an exception in their expert eyes; while other men sent their wives off to water cures for rehabilitation after childbirth, Frank gave Lillian more to-do lists to fill her weeks of postnatal confinement. Correcting galleys for Frank's upcoming books seemed to pass the time better than anything else. *Applied Motion Study* and *Motion Study for the Handicapped* were products of her forced convalescences.

Just months after childbirth, Lillian was eager to attend the first Conference on Scientific Management at the Tuck School of Dartmouth College. Nursing a newborn forced her to keep a low profile, and yet at the final session the chairman asked her to address the audience. Reluctantly, she talked about the "human element" and managed an impromptu case for bringing the study of *human psychology* to bear on systems of efficiency.[24]

The Dartmouth conference was her coming out of sorts. She made such waves among academics that Frank urged her to get back to writing the PhD thesis. There was no time to lose; she had to publish while the iron was hot. He put her on a rigorous writing schedule and hired stenographers to transcribe notes she spoke into a Dictaphone placed near the home office phone. Even as she nursed, she could save time and energy by speaking into the mouthpiece. She finished quickly and rushed the thesis off to her committee at Berkeley, but her submission of "The Psychology of Management" was not nearly as momentous as Frank had hoped. University officials decided not to waive the required year of residence on campus, and her thesis was summarily denied.[25]

The bad news turned out to be the least of her problems. Taylor attributed the advent of motion study to himself during hearings in Washington, while Lillian was ministering to two diphtheric girls at home. Anne recovered, but Mary, her five-year-old, died at the end of January in 1912, the day after Taylor's damning testimony. Frank regrouped by securing a long-term contract at the New England Butt Company and moving the family to Providence, Rhode Island. He welcomed the change of scenery, and his sister was nearby to supervise the children's music training and to lend a hand at home, for Frank Jr., the first Gilbreth boy, was born amid the grieving. Still, the death in Plainfield left permanent scars. Lillian had wanted nothing more than to hold her girl in her arms through the child's final days, but Martha and Frank forbade it, worried about the health of her unborn child. Lillian continued to believe that she could have saved her daughter if given the chance, with motherly love if nothing else.[26] This was the human contingency for which one couldn't plan, and it reinforced her desire to understand the human element. She forever saw efficiency as a noble aim, but not at the expense of an individual's emotional needs. Whether her perspective was the consequence of her biology or her social need to mother, it soon became her lasting imprint on the fields of industrial management and engineering.

"The Human Element": Lillian Gilbreth Feminizes Scientific Efficiency, 1912–24

Providence was a place of new beginnings. Lillian birthed six children there and helped Frank solidify his place in the efficiency movement. Engineers watched closely as husband and wife reorganized the New England Butt Company, introducing micromotion study and one of their most innovative management devices: the "process" or "flow" chart, for visual analysis of production at all its stages. Other clients called on them. A handkerchief manufacturer watched them reduce the movements of workers so significantly that they finished three times more cloth per hour, without added fatigue. Secretaries at the Remington Company learned to perform calisthenics to add to their flexibility, strength, and blood flow; the Gilbreths helped them win the National Typewriting Championships in 1916. Surgeons meanwhile discovered they could remove the therbligs "search," "find," and "select," from their operating repertoires by calling out numbered instruments as nurses placed them squarely in their palms, and major league baseball teams signed up to have their batters, catchers, and pitchers filmed and analyzed. As World War I began, motion study facilitated the reintegration of disabled soldiers and amputees into the civilian workforce.[27]

Taylorites insisted that there was nothing unique about the Gilbreth system, but Frank's patent portfolio grew in Europe and domestically once Taylor died in 1915. Pierce Arrow, Zeiss, Lever Brothers, Eastman Kodak, U.S. Rubber, and Erie Forge Steel were just a sampling of his clients. He printed glossy pamphlets and supplied films to serve as pictorials for his micromotion experiments. "The One Best Way" appeared to be a humanitarian mission rather than a capitalist exploit, since the films included women and disabled workers. Frank erected "Fatigue Museums" that contained badly designed chairs, workbenches, and "instruments of needless torture," underscoring the benevolence of motion study by contrast. Soon the Gilbreths opened a summer school for college teachers of industrial management; Frank was away so often that the running of it fell to Lillian—yet again a case of his tireless promotion and her dogged follow-through.[28]

Frank's business gained from positive publicity about his growing family in the local and national press, where the image of Lillian tending to babies and school lessons humanized his work. More important still, Lillian's knowledge of social science helped him make his definitive mark, for her ideas about the psychology of workers filled his books. In *Motion Study*, for example, he focused on the physiology of workers, but also included their "temperament" and "contentment." It wasn't enough to provide workers with good lighting and ventilation, he told managers. They needed entertainment and a clear understanding of the systems of "reward and punishment." In *Field System* he (and Lillian to an undetermined extent) told managers to cater to workers' needs with "suggestion boxes" and periodicals that stimulated their minds. Lillian developed these ideas further yet, with intricate systems of communication, incentives, fatigue control, and self-regulation in the Berkeley thesis she submitted in 1911, but for three years no publisher dared to print it.[29]

"The Psychology of Management" occupied a precarious place in efficiency literature: it seemed to legitimize the practices of scientific managers and undermine them at the same time. Taylor's disciples had relied on the irrefutability of empirical measurements of time and profits, and yet Lillian surmised that the unquantifiable emotions of individuals crept into their scientific processes. Scientific management was as much process as bottom line, art as much as science. The study of the mind was appropriate in the training of not only teachers and philosophers, she argued, but also engineers, since efficiency ultimately relied on the human worker. One must modify equipment, methods, and materials to make the most out of *him*, she explained. *His* mind "is a controlling factor in his efficiency." Scientific managers "Taylorized" work, but Lillian tried to "tailor" work so that it might fit a man like a glove. Tasks could be surrounded with individualized incentives to accommodate his needs and wants. Suddenly extemporaneous movements were eliminated on the production line, as workers increased output at lower cost to their employers. The benefits felt by capital were shared in her vision; the worker received higher wages but also gained self-esteem as his unskilled labor grew more efficient and even skilled. The gap between the apprenticed worker and the college-trained one diminished, as the relationship between labor and capital turned cooperative. The result was industrial peace.[30]

Lillian's ideas piqued the interest of unionists, social reformers, and academics, but capital industrialists and managers felt threatened by her attention to the human element. When Frank tried to sell her thesis, publishers wrote it off as both archaic and ahead of its time. It didn't help that prominent academic men were starting to write in a similar vein. Harvard psychologist Hugo Munsterberg, most notably, became known as "the father of industrial psychology" after *Psychology and Industrial Efficiency* came out in 1913, and other established scholars jumped on the bandwagon. Desperate to get Lillian's work out in any form, Frank arranged to have parts of her manuscript published serially in *Industrial Engineering* under "L. M. Gilbreth" before Macmillan agreed to publish it in its entirely under the same gender-neutral name. In all the marketing for the book there was no mention that its author was a woman; its reprinting in 1917 and again in 1918 suggests that the strategy worked.[31]

None of this was of much consequence to Lillian, who at the height of the publishing frenzy had delivered an underweight son. But Frank was convinced that Lillian would be above reproach if she had the PhD. He had revisited the issue when they first moved to Providence, but Lillian was lukewarm to the suggestion. The Butt Company contract and the children kept her busy, she told him, but Frank had already been down the street pulling strings with Brown University administrators, who were interested in conferring a degree in "applied management." They'd give Lillian credit for course work already completed and would let her take oral examinations, but she would have to write another dissertation from scratch. "Don't worry, Boss," Frank assured her, "You can see our house from the classrooms. If you see one of our girls climbing out a window, you can run home and catch her before she hits the ground."[32]

It did make all the difference that the university was down the street. Lillian ran home between lectures to nurse the baby before anyone realized she had left. Once the house at 71 Brown Street grew too cramped for the expanding family the Gilbreths moved next door, to a corner lot better equipped with office and lab space for motion study. Lillian watched the children and took notes almost seamlessly once Frank installed Dictaphone equipment on both the first and second floors. "If I haven't but fifteen minutes to dictate," she told a reporter, "I utilize that time"; this was "the secret to the annual book." Every accommodation was made for efficiency; had it not been for the motion studies, the summer school, and the birth of Lillian Jr., she might have taken one year to complete courses instead of two. For most of these months Frank was nowhere to be found; on assignment overseas, he charged his wife with running the office and laboratory and getting the children fluent in German as soon as possible.[33]

It would have been overwhelming had there not been plenty of hired help in the house. Lillian didn't make a meal or scrub a floor; the children were her only persistent chore. There was the domestic supervisor, the cook and handyman, the governess, the maids, and the part-time laundress and hairdresser to lighten the load, and Grandma Martha was still vigorous enough to oversee operations. Time was at a premium, but Lillian was helped by a schedule: two hours for breakfast and the grooming of the children in shifts, two hours to write, a fifteen-minute break with the children before lunch, another hour with children, a half-hour nap, a half hour with the youngest baby, another hour writing, an hour for callers, an hour with the children, a half hour for miscellany, and an hour for dinner before putting the little ones down, helping with homework, and reading bedtime stories. Lillian insisted that the key to keeping focused was her afternoon nap: ten minutes of lying flat on her back did more to eliminate fatigue than ten hours of sleep once overtiredness set in.[34] With this system of efficiency in place, her thesis write-up still went more slowly than Frank would have liked, for Lillian tolerated the children's interruptions and only later instituted a chart to keep them to a minimum. On top of babies and clients, she was also conducting research for *Fatigue Study* and serving on the PTA and the Association of Collegiate Alumni. Frank offered words of encouragement from abroad and kept Lillian regimented with work charts. "I hope that the thesis will not give you nervous prostration," he wrote—hollow words as his pregnant wife feverishly prepared for orals.[35]

Lillian managed to write a four-hundred-page document called "Some Aspects of Eliminating Waste in Teaching," a culmination of hundreds of hours of observation in Providence schools to bring efficiency to the classroom. She gave recommendations for lighting, clothing, ventilation, desks, supplies, and classroom layouts to economize on physical motions, and she advised teachers on how to plan lessons in advance. But again the strength of the thesis was its insight into the human element—what motivates teachers to teach and students to learn. She described what psychologists would later call "the Hawthorne Effect": like the workers involved in her micromotion studies, teachers and students proved eager when they were made active partners in classroom experiments.[36]

As Lillian was finishing her manuscript, her tenth wedding anniversary came and went; Frank could not be Stateside, but he wrote of his undying affection—as well as with instructions for a session at the American Society of Mechanical Engineers (ASME) annual meeting. He returned to Providence after Christmas but was back in Europe in early 1915; he would miss Lillian's graduation in June. She walked in the late spring processional as the lone woman doctor, content in knowing that her older children were to greet her on the campus green afterward. Brown University president W. H. Faunce wrote Frank in admiration of his wife: "I do not know another woman in America who has achieved what she has done in the realm of study, and at the same time fulfilled every duty of motherhood in her constantly enlarging home." Indeed her household continued to enlarge, but not right away: the baby girl Lillian carried under her doctoral robe was stillborn that September. She and Frank said nothing about it to stave off criticism that she was trying to do too much.[37]

Lillian was the first of her peers in scientific management to have the distinction of a doctoral degree. With "PhD" beside her name it now appeared with Frank's on professional papers—more than fifty in total over the following nine years. And yet for all her new access to the profession, her daily existence felt much the same. She nursed babies, transcribed notes, and tended to professional correspondence as Frank sat in hotel rooms pouring out his ideas for the future. "I'm still thinking of the paper you wrote for the Academy of Science. I think that and the Toronto paper and one or two at Dartmouth and some of the old ones printed in *Industrial Engineering* would make a good book for say $1.00 . . . Yes Boss, I see many books that we can put over, easily one a year the rest of our lives and perhaps two."[38]

Late in 1916 Lillian was wrapping up another semester of the Gilbreth Summer School and checking galleys for *Fatigue Study* when Fred, her eighth child, was born. Frank was out of town, as he would be when Dan, the ninth, came thirteen months later. One would think it time to settle in at home, but with U.S. intervention imminent in the war overseas Frank announced that he was leaving to don a military uniform. With no initiation on the part of the army's Engineering Corps, he offered his expertise in motion study. Lillian, as always, let him dream in superlatives—the most successful consultancy, the biggest family, the greatest impact on his country and the efficiency movement. Never would she allow her need for time, space, or physical recovery get in the way of this dreamer's best-laid plans.

And so Lillian held down the fort as "Major Gilbreth" set off for Washington. "If you take care of the kids and answer the mail you will have [done] more than enough," he assured her, but almost immediately he began writing with his fussy requests. She wrote back to assure him that she was getting everything done: she sent the motion models to the Smithsonian and the stills to colleagues, phoned the clients, sent the requested reprints, planned the "Tech talk" she was giving in his stead . . . All was well at home, she reiterated, though the children's interruptions were to blame for any typos on the page. She had dashed up to the school for parent-teacher conferences, taken Anne shopping, attended Ernestine's violin lesson, nursed the baby, put the little ones to bed, and worked in her office as she supervised the older children's homework. She moved the piano into the library and the bookshelves to the parlor. "Now the children have a nice warm place to practise in, and I can inspect typing and piano, and work at my desk all at once."[39]

Rather than cordon off family obligations here and professional ones there, Lillian grew comfortable combining the two, and did it in productive ways. This skill came in handy one month later, when Frank took ill at Fort Sill. She arrived in Oklahoma to find him comatose; immediately she took over, directing his caregivers and instructing the children and domestics through letters sent home. When Frank was moved to Walter Reed Hospital, she shuttled back and forth between Providence and Washington to keep him abreast of the projects she ran in his absence. By August he was discharged to rest with the family in Nantucket. He hobbled on a cane as Lillian ran the household and consultancy and nursed him back to health. She was losing hair and weight, but this was no time to slow down. Gilbreth Incorporated lost money during Frank's convalescence, and it grew clear that they were in need of a bigger house once again. In 1919 the Gilbreths moved to an old, sprawling estate in Montclair, New Jersey, where the schools were good, close associates lived nearby, and Frank would have easy access to his paying New York clients. Lillian made an office out of one of the two ground-floor living rooms and converted the barn into a laboratory with a built-in darkroom.[40]

Modern science was, by definition, not a domestic affair, but for the Gilbreths efficiency demanded the collapse of personal and professional space. This had already been the case in Providence, when the need to remove the tonsils of four of the five eldest children turned into an opportunity to run experiments in surgical time-saving measures right in the house. The Remington contract likewise called for the children to be filmed in typing trials in the room where they normally did their homework. The eldest children served as lab rats for Frank's experiments with colored-coded and Braille keys, and Lillian recorded the typos and times. In the more spacious accommodation in Montclair, Frank outfitted the rooms more completely for motion study. Bathrooms became micromotion labs, as the children groomed to German lessons simultaneously piped into the walls. Bedroom ceilings were painted with Morse code messages and astronomy lessons for absorption during morning and nighttime routines. The children counted in Latin while doing their calisthenics, exercising both body and mind at once.[41]

"Efficiency," Frank Jr. recalled, had become a byword in the Gilbreth household, "the most overworked single word in our vocabularies." The children played with their parents' speed clocks and time exposure equipment and kept their own time as accurately. Lillian did not object when Frank took the children into clients' plants or asked them to stand in for workers in his films. The children turned into consultants as they sat around the movie projector and analyzed the movements of women packing soap flakes. They saw the household bills for the hourly services of repairmen and medical professionals and learned that time is money—don't let family resources go to waste. Everyone chose outfits the night before to wear the next day—a practice arguably *too* efficient to be practical, until the children's studies determined that weather forecasts were accurate 85 percent of the time.[42]

Frank used work charts to plan and regulate the children's activities. Martha's, for example, read "open bed," "dress as far as wash," "wash or bath," "brush teeth," "clean nails," "complete dressing," "hang up night clothes," "piano," "take down wraps," "breakfast," "take school money," "walk to school," "school," "ride home," "wash," "comb hair," "lunch," "walk to school," "school," "ride home," "play," "hang up wraps," "typing," "wash," "supper," "cello," "languages," "report on chart," "undress," "wash or bath," "brush teeth," "lay out clothes," "prayers," "bed." When she completed tasks on time, she marked corresponding squares in blue. Yellow meant completed but late; red meant not completed at all. Frank Jr.s' morning routine from 7:00 to 7:30 didn't allow for a minute of waste: Martha called for him to rise and shine on the hour, and he was to turn into the bathroom by 7:01. By 7:02 he was playing German records as he brushed his teeth and bathed. He weighed himself and recorded it on his weight chart at 7:07, combed his hair, washed the ring out of the tub, started the bath for Bill, and wound up the Graphophone at 7:08. By 7:10 he woke Bill and turned on the French records to listen to while dressing. At 7:16 he shined his shoes, made his bed two minutes later, and straightened the room three minutes after that. By 7:31 he was ready for breakfast—if he was keeping pace.[43]

The children had their "standing order forms," "output charts," "pay charts," "evaluations forms," "telephone charts," and "bath charts,"— even charts recording chest expansion at inhalation and exhalation. The charting grew obsessive but was part of Frank's "One Best Way" to run a household. A journalist for the *Montclair Times* likened the house to a well-run "industrial community," with Frank as the municipal employer, Lillian the manager, and the children their dutiful employees. To understand how literal the analogy became, one needed only witness the children's submission of bids to take on household projects. Frank posted orders for yard work, painting, or repairs, and the child proposing the quickest, cheapest services secured the contract. As if in the factory, each child had an assigned number for routing intra-family memoranda, and Frank installed his "three-point" promotional system, whereby the children were offered incentives to learn, master, and teach household skills at all times. Tasks in the home were broken down precisely to fit the children's physical and mental abilities. Rather than assign a single child to dusting, for example, taller children took the tabletops and high shelves and shorter children the legs and lower shelves. Heights and weights were recorded regularly for the purpose of reassignment. The children took part in these unusual experiments because Dad had instilled the belief that efficiency made them successful people. It was part of his master plan that they skip grades, and most of them did, since they had been taught foreign languages and "mental math" as toddlers. Even the infants were trained to retrieve slippers from marks placed on the closet floor to another set of marks at the foot of his bed.[44]

Such experiments may seem dehumanizing, but Lillian tacitly approved. Watching her children's reaction to them made her sensitive to the psychological needs of workers in the plant, and the lessons served the children, too, when they left home to fend for themselves. Throughout their childhoods, as they prepared statements for weekly family councils, they learned to get points across succinctly for future business meetings. They trained their siblings and received training in return, so that they knew what it was to manage others and accept criticism. They learned to draft their own charts, choosing wording and paper colors that were precise and psychologically pleasing. Lillian insisted that making a fifteen-year-old file income tax reports was too enriching not to do it. The key to compliance was no different from what she espoused in *The Psychology of Management*: provide incentives for the work and provide suggestion boxes that open channels of communication with "management." But most crucial was her behind-the-scenes emphasis on "worker satisfaction." As she watched her children's rigorous routines, she reminded them that they were loved and appreciated for their individual attributes.[45]

This emphasis was hers alone, for Frank was too concerned with a master plan for the group to cultivate individual personalities. He rationalized that the time his children saved could be banked for one-on-one "happiness minutes," but Lillian was the parent cashing them in. "She knew what every one of her individual children wanted, needed, dreaded, and dreamed about," Frank Jr. recalled. "And when a child talked to her, she listened and listened, and made sure she understood." Years later Lillian still advocated for "standardization" in the home. It made sense to maintain uniform sets of buttons, kitchen utensils, and underwear, since that made them cheap and easy to replace. But the practice had to be discarded once it cramped a child's sense of self; in the home, as in the factory, the *individual* was her primary interest.[46]

The Gilbreth children revered their father but recognized that Mom's humanizing perspective was the key to their joint success, both as parents and engineers. It was not coincidence that after Frank's death many of the household systems grew lax, some defunct. Lillian never said much while he lived, but she could see that, in Frank's haste to write papers and graduate children, something had been lost. He could not understand why editors looked so critically at the manuscripts he dashed off, but she knew that they were lacking in art and substance. In business, he conceived the most efficient industrial practices, as she quietly humanized them; and at home, he installed the science she gently improved upon. A reporter noted that, while Frank seemed frantic, his wife "radiated efficiency" without "the slightest evidence of nervous drive or hurry about her. There was a calm and placidity—and femininity" that she found "restful and refreshing." Frank once said that his wife's greatest accomplishment was the *Psychology of Management*, but Lillian disagreed. Babies trumped books, she insisted—always the human element in the end.[47]

**"If the only way to enter a man's field was through the
kitchen door, that's the way she'd enter.": 1924–**

Grandma Martha outlived Aunt Kit and died in Montclair at eighty-six years of age. Jack, baby number ten, was months old, and number eleven was on the way. Lillian mourned the loss of Martha's help at home. To afford their mortgage, she had let go of much of the personnel they had had in Providence, but Frank was confident that with his procedures in place, she would not have to relinquish her professional endeavors. In fact after the birth of Jane, baby twelve, the Gilbreths had planned a seven-week lecture tour in Europe. Lillian looked forward to giving the keynote address to the British Society of Women Engineers, but this time an emergency hysterectomy, not a birth, kept her grounded. The trip with Frank was never to be, for he died the following June. With blinders on, Lillian proceeded stoically down Curie's anti-natural path: "The rest is bringing up the children, and proving I deserved all Frank gave me—and that was everything."[49]

Lillian was herculean in her efforts to parent and pay the bills after Frank died. She got her kids ready for school, supervised menus and budgets, sewed buttons, wrote the kids at college, read stories, and helped with homework—in addition to ten-hour days in the office and laboratory. Her older children said that she rarely missed a class play, a commencement, or a "Be Your Child" day at the Montclair schools (for which she hired a cab to chauffeur her to multiple classrooms at the elementary, junior high, and high schools for the day). But any woman who has ever tried to balance work and family knows that no path is walked without sacrifice. The family log indicates that there were years when she traveled more than she was home. When she went abroad the children made a ritual of waiting for the mailman to arrive with her letters. On her return from the World Power Conference in mid-1930, she could not go home immediately: Martha picked her up and drove her to a consulting job in Manhattan and then to catch a train to the Society of Industrial Engineers (SIE) meeting in Washington. Jane, who was too young to know her father, felt that she suffered the most. Her older siblings had enjoyed family vacations with both parents in Nantucket every summer, yet she was shuttled off to summer camp so Mom could go abroad "for the millionth time."[50] The astute observer of people, Lillian had to be aware of her daughter's resentment; yet as a single parent and breadwinner, she had to sacrifice the human element from time to time.

The day after Frank's death, Lillian wrote the president of SIE asking, in light of recent events, if she could speak at the Power Conference in London. The answer was an unequivocal yes, for colleagues thought the gesture not unlike Marie Curie's completing Pierre's lectures when he died in 1906: this was the loving tribute of a devoted wife, not the trespass of an ambitious woman. She gave her lecture in London, completed Frank's contracts in Holland, and received a warm reception in Prague. After she returned she published her favorable impressions of the trip in *American Machinist* and then fulfilled Frank's speaking engagements on college campuses throughout the country. In 1925 SIE published her biography of Frank, *The Quest of the One Best Way*.[51] Immediately after Frank's death, the profession appeared to embrace her.

But when the mourning period was over, like Marie Curie, Lillian discovered that the tributes to her husband were not a sign of her own acceptance. Within the year her three biggest clients canceled contracts or refused to renew. She paid visits to Winchester Laundries, Filene's, and other standbys, but they, too, were reluctant to rely on her for their managerial needs. Close colleagues offered her positions in their firms, which she politely declined in favor of keeping Frank's business afloat. To expand her contacts she turned to SIE, where, although women were not allowed to become regular members of the society, they could attend meetings, making it the most permissive of the national organizations. As an honorary member Lillian was able to become chair of its Fatigue Elimination Committee in 1926. The position brought her speaking engagements, as did several of Frank's university contacts, but the attached honoraria rarely covered the expense of travel. Frank had grown accustomed to first-class accommodation, but his wife booked upper births on trains and stayed with friends in college towns.[52]

The forces of institutional chauvinism pushed her further out of industrial engineering—and she wasn't alone. By 1938 the list of more than eighteen hundred women in *American Men of Science* included only five women engineers in any subfield. When Frank was alive she had been insulated from discriminatory practices; as a widow, however, she felt the wrath of a profession that relied on its virile image for definition. The engineer conjured in the American mind was rugged and outdoorsy, seemingly blue-collar in his affiliations. Shop floor workers and managers had always seen Frank as one of them, when he rolled up his sleeves and smoked his cigars. Needless to say, they could not view Lillian in the same manner, nor could other professional engineers. She was turned away from a dinner at the University Club in New York even though she was an invited guest: according to building regulations, she, like any woman, was not allowed on the premises. When a committee she was on convened over breakfast at the Engineers Club, she had to eat elsewhere, and colleagues never considered relocating to another venue. Men introduced her as Mrs. Frank Gilbreth, although men with identical credentials were referred to as "Dr." this or that.[53] Her reaction was always gracious; she saw no point in making waves.

Lillian managed to get honorary membership in the American Management Association and sought the same in the ASME, but the odds were against her. Kate Gleason, the only other female member, had courted such disfavor that another woman's candidacy seemed unlikely. Gleason was thirteen years Lillian's senior and had much in common with her, since she, too, had learned her trade "in the family," inventing worm gears at her father's machine tool company. But her single status troubled male peers, for without a husband, there was no one to whom her accomplishments could be attributed. Lillian found herself in the same position after 1924. A loyal few lobbied behind the scenes to see that she got an honorary membership in the ASME, but it was clear that she would have to refocus her energies in directions that male colleagues thought more appropriate to women. Gleason had left tool making to build low-cost housing; Lillian, too, turned to civic projects, assisting the down-and-out during the Great Depression. Under President Herbert Hoover she headed the women's division of the Emergency Committee for Employment and served on the Organization of Unemployment Relief.[54]

It was not unusual to see a woman engaged in social welfare work, for American women had worked inside church groups for generations and inside government agencies for decades. Rather than fight cultural assumptions about her strengths and proclivities, Lillian allowed them to buttress her position of authority in areas where her persona as mother of twelve and nurturer of the human element would be valued. If she could not work in the industrial plant, she'd take on the kitchen, the classroom, or the retail store. When mechanical engineers remained hostile, she gravitated to growing communities of management consultants and industrial psychologists. And when she couldn't get hired to install the Gilbreth system in plants, she decided to teach it to those who could. The vice president of Johnson & Johnson proposed that she open another school of motion study for his managers, and she embraced the idea since it allowed her to stay close to home. Charging students a thousand dollars for the course, she managed to pay the family's bills and college tuitions. Over six years she attracted managers from General Electric, Borden Milk, even international firms.[55]

The course was supposed to advance Frank's motion study, but her added views of plant psychology kept the students coming. Since her pioneering work had appeared, the number of organizations and publications dedicated to the subject had expanded. She became a frequent editor for *Industrial Psychology* and an instructor for newly formed institutes in the field. Increasingly, however, clients sought her out as an expert on the *woman* worker in particular, and she was willing to remake herself accordingly. In the early Depression heavy industrial sectors staffed predominantly by men were hardest hit. Until then, most middle-class women who worked were single. Now for the first time in American history significant numbers of married women entered light industry and pink-collar services to compensate for their husbands' lack of gainful employment. Their growing numbers reopened debates about protective legislation and women's proper place at home. Women were becoming integral contributors to the family wage, but their responsibilities as homemakers only intensified: pinching pennies and stretching leftovers were keys to weathering an economic decline. Their double burden as workers and homemakers was unprecedented, yet Lillian had lived it all along; there could be no better authority than she on women's work and efficiency, both in and outside the home.

Lillian combined her expertise on waged work with studies of domestic economy that women had pioneered since the late nineteenth century. Socialist feminists such as Charlotte Perkins Gilman had experimented with the idea of communal living arrangements and technologies to lighten the load at home and free women to pursue paying careers. These efforts had converged with those of Ellen Swallow Richards, Helen Campbell, and other women who reinvented themselves as domestic scientists when the academy refused them entrance to other science fields. By 1910 they had christened their field "home economics" and had created the American Home Economics Association

(AHEA). Such popular writers as *Ladies Home Journal* editor Christine Frederick helped sell books and magazines through this idea of alchemy in the home in the 1910s, and the Gilbreths, too, had begun to think of ways to make the homemaker self-sufficient.[56] Home economists from Columbia's Teachers College had come to their home to perform motion studies of bed making, and Lillian had adopted a cross-sectioned kitchen for more studies. The AHEA's response was enthusiastic, but in the Gilbreth household prestigious industrial clients had taken priority, since they helped to pay the bills.[57]

Ironically, once Lillian was alone, domestic efficiency was nearly the only form of scientific management for which clients sought her expertise. In 1927 Mary Dillon, president of the Brooklyn Gas Company, asked her to develop the prototypical "Kitchen Practical" for the Women's Exposition of 1929. The working space in this model room was circular—from the refrigerator, to the kitchen cabinets, the stove, the sink, and the serving table—and its diameter was the distance between the homemaker's outstretched arms, shoulder to fingertips. For optimal convenience, the service table had wheels so it could move anywhere in the room, and the heights of surfaces were adjusted to eliminate fatigue. Stools brought a child's workspace level with his or her mother's, for in this kitchen cooking became a *family* affair. In the promotional literature a man in business attire donned an apron and cooked alongside his wife and child with a look of contented bliss. Lillian insisted the kitchen be practical, but attractive enough to inspire creativity. This was a space where science integrated with domesticity as well as art to turn homemaking into a fulfilling endeavor for the American homemaker.[58]

Indeed Lillian believed that scientific management and homemaking were not antithetical. If installed correctly, scientific principles could eliminate physical fatigue, the psychological drudgery of housework, and the low self-esteem of the homemaker. Lillian gave many radio addresses touting the merits of motion study in the home, and she received back-to-back contracts to write *The Home-maker and Her Job* and *Living with Our Children* in 1927 and 1928, respectively. From canning baby food to designing workspace, Lillian's books advised on "the One Best Way" to run a household. Since readers were unlikely to have micromotion equipment, she encouraged them to design homemade experiments. Their children, for example, could reproduce a cyclegraph by retracing Mom's movements with a ball of string and pinning it every time she changed direction. Why not make simo charts to establish the best posture for washing dishes or count the therbligs to bake a cake?[59]

The methods Frank had applied to bricklayers and baseball players lent gravity to the housework of American women and turned them into experts in their own right. Lillian reminded women that, using both the human insights of psychologists and the analytic skills of engineers to run their homes, they were, in essence, managers of human beings and material resources. She also reminded them that efficiency was never an end in itself: it allowed them more time to spend with their children and in other endeavors that might afford them pleasure. If these activities took them outside the home, then so be it. Lillian's ideal homemaker was not a glorified one so much as a woman fulfilled. The distinction she made between the housekeeper and the homemaker depended on the degree to which an equilibrium between housework and career work, work for subsistence and work for vitality, work for others and work for oneself had been achieved. Housekeeping was science, but homemaking was "housekeeping plus"—the art of directing science toward the most creative ends.[60]

Lillian wrote her domestic books as she took on contracts to train secretaries and reformatory matrons because she saw the relationship between home and workplace as parallel and reciprocal. Eventually she merged her knowledge of both realms into a contract for Macy's, a department store where domestic consumers and female workers literally shared common ground. Eugenia Lies, a student in her motion studies course, was head of the Planning Department and invited Lillian's students to make a case study out of the Manhattan store, specifically in the problem area of the "tube room," where centralized cashiering took place. Lies convinced her bosses that Lillian's female sensibilities could provide solutions for the store's sales and human resource problems, for Lillian understood worker psychology and, like 85 percent of department store clientele, ran a household of her own. Management agreed that Lillian would be likely to understand and anticipate the spending practices of female shoppers, but they were dubious about her ability to understand the needs of the female sales force and cashiers. For more insight Lillian spent a summer working in store departments, getting into the heads of disgruntled women on the floor.[61]

No stone went unturned during the three years she spent tightening operations. She changed light fixtures to reduce eye fatigue, repadded walls to reduce noise in the tube room, determined the fewest therbligs for working the cashier desk, and did away with duplicate recordings of sales checks. She implemented procedures to reduce counting errors and the amount of time customers waited for change. One result was happier customers with more time and interest in spending money in a pleasing store, newly remodeled and easy to navigate. More efficient cashiers could ease the pressure on sales clerks, who needed to make their daily quotas quickly. The mainly female employees reaped rewards as profits were passed down in the form of time off, cash, and promotional incentives. Lillian revamped managerial practices, too, creating better systems of posting and filing of employee records, and opening channels of communication between managers and salesclerks. As she had insisted in *The Psychology of Management*, the individual worker practically managed herself when she was consulted about changes. Her fatigue decreased when jobs were fitted to her physical and psychological needs.[62]

By 1932 Lillian was speaking to members of the National Retail Dry Goods Association about her new expertise, "What the Customer Wants"—particularly the woman consumer. Her work at Macy's attracted Sears Roebuck, and Johnson & Johnson hired her to develop and market Modess sanitary napkins. Men with business degrees were stymied about how to amass consumer data on a product that, until only years before, women had privately made themselves. Kimberly Clark had tried to market Kotex pads, the first commercial feminine hygiene product, but no marketing team had mastered the art of selling them to women, who were too embarrassed to offer product feedback. Because Lillian and her research team were female, she got candid responses from consumers: They wanted greater comfort, protection, and inconspicuousness in a product they could discreetly obtain and throw away. Back in Montclair she created her own research lab, flushing, submerging, and pulling apart products already on the market and observing women's reactions to the size and shape of the boxes they came in. Lillian seized on the rare fact that a corporate bottom line relied on a thorough study of women's bodies; her report to Johnson & Johnson included exhaustive data on women's cycles and their attitudes about menstruation.[63]

Lucy Maltby, a contemporary, had developed Pyrex ovenware for Corning Glass Works. Lillian, too, had found a way to make being female seemingly enhance her expertise rather than diminish it. By 1931 the domestic guru was asked to promote "the Management Desk," a streamlined piece of furniture equipped with clock, adding machine, radio, telephone, child reference books, and charts for the organization of domestic chores.[64] The desk, as well as her prototypical kitchens, seemed to yoke women to the home by systematizing their domestic operations. But one could also argue that they had a liberating effect, when they saved homemakers time to enter the masculine professions. One of her designs was actually a kitchenette for a two-career household that became the blueprint for one she created for her daughter Ernestine, who had married and worked as a buyer for Macy's. Over the following three decades Lillian's ideas improved women's lives in a number of ways. Using motion studies, she taught wheelchair-bound women how to make beds and supper, and she created specially rigged kitchens for women debilitated by heart disease. She teamed up with home economists to publish floor plans that American women could purchase and adjust to any need—be it validation of their domesticity or efficiency to work outside the home.[65]

Regardless of the message women might read from her designs, Lillian's image was the best tool to market them; she represented glorified motherhood and ambitious careerism all at once. Exhibitors of her Kitchen Practical handed out coffee cake recipes to passersby that they claimed were hers—an ironic scene, given her actual experience in the kitchen. "Stoves burned her, ice picks stabbed her, graters skinned her, and paring knives cut her," Frank Jr. recalled. This was a woman who concocted a creation that her children privately referred to as "Dog Vomit on toast." When cameramen came to Montclair to shoot promotional films for her kitchen designs, she quickly had to remodel her own. And yet her remaking as a domestic guru was complete. Frank had invented an adjustable scaffold; she, a foot-pedaled kitchen trashcan. He had reduced motions for bricklaying, and she reduced the effort required to make the breakfast coffee. She developed electric stoves, refrigerators, and washing machines and described how to mix a cake, bake it, and clean up the dishes in just a few dozen steps—even if she had never taken them herself.[66]

Her popular image was fictitious to a degree, and yet she had been efficient enough to bank her happiness minutes so that she might realize a career on her terms. Engineers on shop room floors did not necessarily know who she was, but journalists for homemaking magazines referred to her in the same sentence as June Cleaver, Betty Crocker, and Dr. Spock. Americans admired her maternal patriotism, first as a consultant to relief agencies during the Depression, and then as an executive member of the Girl Scouts, the War Manpower Commission, and women's army and naval auxiliary boards (WACS and WAVES) during World War II. Agencies under Hoover, Roosevelt, Truman, Eisenhower, Kennedy, and Johnson used her motion studies to bolster civil defense, increase war production, rehabilitate the disabled, and care for the aged. Walt Disney made a Technicolor training film of her process charts, and her portrait appeared on postage stamps. In 1948 the American Women's Association named her Woman of the Year, and *Cheaper by the Dozen* made it to bookstands, increasing her celebrity status even more.[67]

Lillian treasured her life and work as a wife and mother, which her children glorified for postwar readers and filmgoers, but the truth was more complicated. She had also longed to engage in professional and intellectual pursuits that were seemingly in conflict with domesticity. She made working life acceptable by becoming a domestic icon even as she transgressed the stereotype. Her ideas for home efficiency did not appear to disrupt the notion that homemakers were female, and yet as early as the 1920s and 1930s she had been describing the work/family balance as a dilemma for men and women both. Her assessment of Frank's contributions to her own home may have been generously distorted, but she insisted that scientific management had equalized her marriage, just as it had democratized the shop room floor. As domestic journals of the 1950s told women to prepare labor-intensive meals to meet expectations of perfection at home, Lillian questioned the merits of such martyrdom, much as Betty Friedan did later in *The Feminine Mystique*. There was nothing virtuous about servicing others at the expense of oneself, she told homemakers. Drive-throughs, mail order, ready-made cakes, laundry services, and other modern conveniences were not enemies but possibly keys to a more balanced life. Lillian's own balance of home and work had taught her not to judge, but to give women the tools to work out the best balance for themselves.[68]

Lillian wanted women to feel fulfilled as homemakers, but that also meant turning her attention to where women made wages and bought their domestic goods. Wherever women were, she lent them legitimacy by systematizing their operations. To call her subversive may be at odds with what she intended, but it's hard to ignore the contradiction she embodied throughout her life. The press referred to her as the "First Lady of Engineering," but she was also honored in New Jersey as "Mother of the Year," suggesting that there was no consensus about which of her virtues to extol: her professionalism or her maternal excellence. Lillian didn't think it an either/or proposition, and indeed members of the Industrial Management Society seemed to agree when it named her "Mother of the Century" in 1959. Mills College president Lyn White thought her life "the biological phenomenon of this century," but Lillian believed that anything was possible with a scientific plan in place.[69]

By 1960 Lillian boasted membership in the Society of Women Engineers, and yet engineering remained the most male of scientific professions. Women made up 4.2 percent of the nation's physicists, 8.6 percent of its chemists, and 26.7 percent of its biologists, but it would take another decade before they made up more than 1 percent of engineers. The few women who dared to enter the field in the early twentieth century—women such as Kate Gleason, Bertha Lamme, Edith Clarke, and Ellen Swallow Richards—were greeted with hostility and settled often for projects that male peers refused. Lillian's marginalized status in the field may have afforded her the best perspective on engineering problems in the end. She had nothing to lose by rejecting the rules of scientific managers, and so she redefined and expanded them. Sound science, she declared, was in the home, the human element, and the culturally female sphere.[70]

Through all her years of hardship, Lillian had sworn that all her children would go to college, and this indeed came to pass.[71] It was not a coincidence that several went to colleges where she had already been given honorary degrees, for ultimately she received more than twenty. Rutgers awarded her a full-fledged Doctorate of Engineering in 1929, and six years later Purdue University broke all gender protocol by making her the first and only full-time female faculty member as professor of management in its School of Mechanical Engineering. She divided her time between motion study in the Department of Industrial Engineering and teaching in industrial psychology and home economics; she became the official "consultant on careers for women," but also enjoyed a week off every month to return to Montclair for family visits. Once she had fallen prey to rigid boundaries between home life, engineering disciplines, and academic and commercial work; in her sixties, she moved between these realms at will.[72]

As a young widow Lillian had carried an oversized pocketbook that contained a hodgepodge of items on any given day: sometimes a shawl she was knitting or socks she was darning, drafts of speeches, back issues of *Iron Age* magazine, and a notebook in which she jotted down reminders about work and her children. Her children noted the irony of there being "never anything very efficient about Mother's pocketbook."[73] Perhaps it is less ironic than emblematic of Lillian and her life. The pocketbook always contained projects—both personal and professional—that she was in the midst of completing placed indiscriminately next to each other. She pulled them out and worked on them whenever and wherever she could spare a minute, not in places and moments specially designated for them. It was not unusual in her younger years to find her feeding a baby as she drafted a scientific paper, or in later years crocheting as she was introduced at a professional gathering. When her kids became adults she continued to write them daily, often from the planes, trains, and taxis shuttling her from one speaking engagement to the next. One might think her a victim of circumstance—a woman who hadn't the luxury of drawing boundaries between her science and domesticity. Or one can see her as the hero of her own story—a scientist whose identity took on expanded meaning. Versatility and efficiency were the keys to her success. Where professional scientists have idealized a separation between worlds, Lillian Gilbreth brought their permeability to light and revealed endless possibilities.

Notes

FG: Frank Gilbreth
GC: Frank and Lillian Gilbreth Collection at Purdue University, West Lafayette, IN, selected papers (Cleveland, OH: Micro Photo Division, Bell and Howell, 1976)
LG: Lillian Gilbreth
LMG: Lillian Moller Gilbreth Papers in Sophia Smith Collection, Smith College, Northampton, MA

1. Brochure for the Kitchen Practical, 1931, Box 14, folder 12, LMG.
2. Lillian Gilbreth, *As I Remember: An Autobiography* (Norcross, GA: Engineering and Management Press, 1998), 190–92.
3. Frank Gilbreth Jr., *Time Out for Happiness* (New York: Thomas Y. Crowell, 1970), 156–57; Edna Yost, *Frank and Lillian Gilbreth: Partners for Life* (New Brunswick, NJ: Rutgers University Press, 1949), 308.

4. LG to Annie Moller, July 6, 1924, Box 11, folder 9, LMG.

5. Margaret Ellen Hawley, "The Life of Frank B. Gilbreth and His Contributions to the Science of Management" (master's thesis, University of California, 1925), 201 (microfilmed on reel 1, GC); "Family Log," entries from June 14, 1924 to June 22, 1924, Box 3, folder 2, LMG.

6. Jane F. Levey, "Imagining the Family in Postwar Popular Culture: The Case of *The Egg and I* and *Cheaper by the Dozen*," *Journal of Women's History*, 13, no. 3 (2001): 125–50.

7. Frank B. Gilbreth Jr. and Ernestine Gilbreth Carey, *Cheaper by the Dozen* (New York: First Perennial Classics, 2002), 205 (the original was published by T. Y. Crowell in 1948).

8. Gilbreth, *As I Remember*, 52–80; Gilbreth, *Time Out for Happiness*, 30–32; Jane Lancaster, *Making Time: Lillian Moller Gilbreth—a Life Beyond "Cheaper by the Dozen"* (Boston: Northeastern University Press, 2004), 21–64.

9. Lillian Gilbreth, *The Quest of the One Best Way: A Sketch of the Life of Frank Bunker Gilbreth* (Easton, PA: Hive, 1973), 13–21 (the original was published by the Society of Industrial Engineers in 1925); "Wife Will Carry on His Work," *Montclair Times*, June 28, 1924, reel 4, GC; Robert Kangiel, *The One Best Way: Frederick Winslow Taylor and the Enigma of Efficiency* (New York: Viking, 1997), 415.

10. Yost, *Frank and Lillian Gilbreth*, 25, 60; Gilbreth, *As I Remember*, 102–3; Gilbreth, *Time Out for Happiness*, 89, 96–97; Lillian Gilbreth, *Living with Our Children* (New York: Norton, 1928), 4–5, 33–34.

11. Gilbreth, *Quest of the One Best Way*, 25.

12. Gilbreth, *Time Out for Happiness*, 170; Gilbreth, *As I Remember*, 107; Typed reminiscence of Ernestine Gilbreth Carey, 1960, Box 3, folder 4, LMG; Gilbreth and Carey, *Cheaper by the Dozen*, 127.

13. Lancaster, *Making Time*, 92; Gilbreth, *Quest of the One Best Way*, 23; Gilbreth, *As I Remember*, 103–7; Gilbreth, *Time Out for Happiness*, 6, 69, 88; Yost, *Frank and Lillian Gilbreth*, 123–25, 149.

14. Gilbreth, *Time Out for Happiness*, 92.

15. Kangiel, *The One Best Way*; Brian Charles Price, "One Best Way: Frank and Lillian Gilbreth's Transformation of Scientific Management, 1885–1940" (PhD thesis, Purdue University, 1987), 7–8.

16. Frank B. Gilbreth, *Primer of Scientific Management* (New York: D. Van Nostrand, 1911). The publisher would not put Lillian's name on the cover. See Price, "One Best Way," 140–41; Kangiel, *The One Best Way*, 414–16.

17. Lancaster, *Making Time*, 111–14.

18. Price, "One Best Way," 288–91; Yost, *Frank and Lillian Gilbreth*, 156–62, 192–215; Gilbreth, *Quest of the One Best Way*, 35.

19. Frank Gilbreth and Lillian Gilbreth, *Applied Motion Study*, in *The Writings of the Gilbreths*, ed. William R. Spriegel and Clark E. Myers (Homewood, IL: Richard D. Irwin, 1953), 220–31 (*Applied Motion Study* was originally published by Sturgis and Walton in 1917); Sharon Corwin, "Picturing Efficiency: Precisionism, Scientific Management, and the Effacement of Labor," *Representations*, 84 (2004), 139–47; Brian Price, "Frank and Lillian Gilbreth and the Manufacture and Marketing of Motion Study, 1908–1924," *Business and Economic History*, 2nd ser., 18 (1989): 91; Price, "One Best Way," 224–37.

20. Brian Price argues that the Gilbreths were more successful at publicizing motion study as the humane alternative than in actually quelling tensions between workers and managers. Contemporaries such as Robert Hoxie argued that they achieved little more than "benevolent despotism" in the industrial plant. See Price, "One Best Way," 7–8, 334.

21. Gilbreth, *Time Out for Happiness*, 180; Gilbreth, *As I Remember*, 106–7, 113, 118, 144;

Lancaster, *Making Time*, 112; Yost, *Frank and Lillian*, 193; Price, "Frank and Lillian," 92–93; FG to LG, October 18, 1914, GC.

22. FG to LG, August 30, 1906; October 1, 1914; November 15, 1914; August 31, 1914, reel 3, GC.

23. The best known of the medical experts to take this view was Edward Clarke. See *Sex in Education; Or, a Fair Chance for the Girls* (Boston: Osgood, 1873); Rosalind Rosenberg, *Beyond Separate Spheres: Intellectual Roots of Modern Feminism* (New Haven, CT: Yale University Press, 1982), 5–13; Cynthia Eagle Russett, *Sexual Science: The Victorian Construction of Womanhood* (Cambridge, MA: Harvard University Press, 1989), 116–19.

24. Yost, *Frank and Lillian Gilbreth*, 193–94.

25. Lancaster, *Making Time*, 118.

26. Yost, *Frank and Lillian Gilbreth*, 208, 213; Typed reminiscence of Ernestine Gilbreth Carey, (1960); Gilbreth, *As I Remember*, 119.

27. "Movies to Help Baseball Players Economize Force," *New York Tribune*, June, 1913, reel 4, GC; Gilbreth, *Time Out for Happiness*, 128, 140–41, 148–49, 154; Gilbreth, *As I Remember*, 128; Gilbreth, *Applied Motion Study*, 220; Price, "One Best Way," 374.

28. Price, "Frank and Lillian Gilbreth," 88, 92; FG to LG, October 2, 1914, reel 3, GC; Yost, *Frank and Lillian Gilbreth*, 250.

29. Frank Gilbreth, *Motion Study*, in *Writings of the Gilbreths*, 152.

30. L. M. Gilbreth, *The Psychology of Management: The Function of the Mind in Determining, Teaching, and Installing Methods of Least Waste* (New York: Macmillan, 1921), 1–3, 18–19 (the original was published by Sturgis and Walton in 1914).

31. Gilbreth, *Quest of the One Best Way*, 36; Gilbreth, *As I Remember*, 120; Yost, *Frank and Lillian Gilbreth*, 213.

32. FG to LG, October 12, 1914, reel 3, GC; Gilbreth, *Time Out for Happiness*, 125; Lancaster, *Making Time*, 127.

33. Mayme Ober Peak, "She Conquers Fatigue—Woman's Greatest Enemy," *Beautiful Womanhood*, February 1923, Box 3, folder 18, LMG; Gilbreth, *As I Remember*, 121–32; Yost, *Frank and Lillian Gilbreth*, 253; Lancaster, *Making Time*, 131.

34. "Instruction Card," typed schedule [1912], Box 4, folder 2, LMG; "Mrs. Gilbreth Gives Formula for Happy Home," reel 4, GC; Peak, "She Conquers Fatigue."

35. Gilbreth, *As I Remember*, 122; Yost, *Frank and Lillian Gilbreth*, 228; Gilbreth, *Living with Our Children*, 230; Lancaster, *Making Time*, 130–31, 222; FG to LG, April 5, 1915, reel 3, GC.

36. Lillian M. Gilbreth, "Some Aspects of Eliminating Waste In Teaching" (PhD dissertation, Brown University, 1915); Lancaster, *Making Time*, 153–57.

37. FG to LG, October 14, 1914; October 18, 1914; April 5, 1915; W. H. Faunce to Frank B. Gilbreth, June 6, 1921, reel 3, GC; Lancaster, *Making Time*, 161–62.

38. FG to LG, April 29, 1915; May 13, 1915, reel 3, GC.

39. FG to LG, January 7, 1918; January 9, 1918; LG to FG, January 7, 1918; January 8, 1918; [2 letters, January 1918], n. d.; January 18, 1918; January 31, 1918, reel 3, GC.

40. Gilbreth, *As I Remember*, 152–61; Gilbreth, *Time Out for Happiness*, 168; Lancaster, *Making Time*, 184.

41. "Mother's notes on Typewriting-training," November 24, 1916, Box 3, folder 7, LMG; Elizabeth Ellam, "Gilbreth Nantucket Laboratory Most Interesting Place," *Nantucket Inquirer and Mirror*, September 8, 1923; LG to FG, January 29, 1918, reel 3, GC; Gilbreth, *As I Remember*, 147; Yost, *Frank and Lillian Gilbreth*, 269–70.

42. Gilbreth, *Time Out for Happiness*, 17, 132; Gilbreth, *As I Remember*, 169; Gilbreth, *Living with Our Children*, 173–79; Hawley, "The Life of Frank B. Gilbreth," 113–6.

43. "Daily Schedule of Ernestine Gilbreth" (Martha's schedule included), n. d., Box 3, folder 6, LMG; Hawley, "The Life of Frank B. Gilbreth," 113; Gilbreth, *Time Out for Happiness*, 147; Gilbreth and Carey, *Cheaper by the Dozen*, 2.

44. "Family Log," March 1, 1923; "Wife Will Carry on His Work"; Hawley, "The Life of Frank B. Gilbreth," 114; Frank B. Gilbreth Jr. and Ernestine Gilbreth Carey, *Belles on Their Toes* (New York: First Perennial, 2003), 2, 11–12 (the original was published by T. Y. Crowell in 1950). Gilbreth, *As I Remember*, 110; Yost, *Frank and Lillian Gilbreth*, 178; FG to LG, January 25, 1918, reel 3, GC; Gilbreth, *Time Out for Happiness*, 129.

45. Gilbreth, *Living with Our Children*, 227–38.

46. Gilbreth, *Time Out for Happiness*, 191; Lillian M. Gilbreth, *The Home-Maker and Her Job* (New York: D. Appleton-Century, 1936), 102 (the original was published in 1927).

47. Gilbreth and Carey, *Belles on Their Toes*, 200; Yost, *Frank and Lillian Gilbreth*, 294; Peak, "She Conquers Fatigue"; Ellam, "Gilbreth Nantucket Laboratory Most Interesting Place."

48. Gilbreth and Carey, *Belles on Their Toes*, 100.

49. FG to LG, July 28, 1922; July 31, 1922, reel 3, GC; LG to Minnie Bunker, July 6, 1924, Box 7, folder 12, LMG; Yost, *Frank and Lillian Gilbreth*, 306; Laurel Graham, *Managing on Her Own: Dr. Lillian Gilbreth and Women's Work in the Interwar Era* (Norcross, GA: Engineering and Management Press, 1998), 86–92.

50. Gilbreth and Carey, *Belles on Their Toes*, 57, 88–89, 131; Gilbreth, *Time Out for Happiness*, 190; "Family Log" entries, 1929–1930; March 26, 1934.

51. Hawley, "Frank B. Gilbreth," 202; Graham, *Managing on Her Own*, 90–92.

52. Gilbreth, *As I Remember*, 201–202; Gilbreth, *Time Out for Happiness*, 169, 187, 196–97.

53. Alice Rossi, "Barriers to the Career Choice of Engineering, Medicine, or Science Among American Women," in *Women and the Scientific Professions: The MIT Symposium on American Women in Science and Engineering*, ed. Jacquelyn A. Mattfeld and Carol G. Van Aken (Cambridge, MA: MIT Press, 1965), 97; Lancaster, *Making Time*, 228; Gilbreth, *Time Out for Happiness*, 188–89; Yost, *Frank and Lillian Gilbreth*, 356–57.

54. Martha Moore Trescott, "Women in the Intellectual Development of Engineering: A Study in Persistence and Systems Thought," in *Women of Science: Righting the Record*, ed. G. Kass Simon and Patricia Farnes (Bloomington, IN: Indiana University Press, 1990), 168; Yost, *Frank and Lillian Gilbreth*, 333–34; Graham, *Managing on Her Own*, 224–25.

55. Gilbreth, *As I Remember*, 202–17; Gilbreth, *Time Out for Happiness*, 203–5; Graham, *Managing on Her Own*, 95–100; Hawley, "Frank B. Gilbreth," 202–4.

56. Christine Frederick, *The New Housekeeping: Efficiency Studies in Home Management* (Garden City, NY: Doubleday, 1913); *Household Engineering: Scientific Management in the Home* (Chicago: American School of Home Economics, 1920); Sarah Stage and Virginia B. Vicenti, eds., *Rethinking Home Economics: Women and the History of a Profession* (Ithaca: Cornell University Press, 1997).

57. See, for example, Lillian Gilbreth, "Fatigue Study and the Home," *Society of Industrial Engineers Proceedings*, April 1921, 33–39; Graham, *Managing on Her Own*, 155–64.

58. "The Kitchen Practical Designed for the Brooklyn Borough Gas Company by Dr. Lillian M. Gilbreth"; "Kitchen Practical: The Story of an Experiment," 1931, Box 14, folder 12, LMG.

59. Lillian Gilbreth, "Is Your Home a Hazard?" Radio talk from "America's Little House," February 19, 1935, reel 3, GC; Gilbreth, *The Home-Maker and Her Job*, 21, 92, 96; Lillian Gilbreth, Orpha Mae Thomas, and Eleanor Clymer, *Management in the Home: Happier Living Through Saving Time and Energy* (New York: Dodd, Mead, 1955), v.

60. Gilbreth, *The Home-Maker and Her Job*, 20.

61. Price, "One Best Way," 619–20; Gilbreth, *As I Remember*, 204–5.

62. Graham, *Managing on Her Own*, 118–41.

63. Yost, *Frank and Lillian Gilbreth*, 321; Graham, *Managing on Her Own*, 218–21.

64. Regina Lee Blaszczyk, "'Where Mrs. Homemaker Is Never Forgotten': Lucy Maltby and Home Economics at Corning Glass Works, 1929–1965," in *Rethinking Home Economics*, 163–80; "A Modern Aid in the Solving of Home Management Problems," 1931, Box 14, folder 11, LMG.

65. "Heart of the Home," 1948–1949, Box 14, folder 13, LMG; Graham, *Managing on Her Own*, 182; Gilbreth, *Time Our for Happiness*, 214–15.

66. Gilbreth and Carey, *Belles on Their Toes*, 100–3; Graham, *Managing on Her Own*, 182–83; Gilbreth, Thomas, and Clymer, *Management in the Home*, 78–81.

67. Edna Yost and Lillian Gilbreth, *Normal Lives for the Disabled* (New York: Macmillan, 1944); Gilbreth and Carey, *Belles on Their Toes*, 218; Gilbreth, *Time Out for Happiness*, 219, 223, 231; Yost, *Frank and Lillian Gilbreth*, 336n.

68. "Man's Place Is in the Home," *Philadelphia Public Ledger*, January 31, 1932; Gilbreth, Thomas, and Clymer, *Management in the Home*, 47; Peak, "She Conquers Fatigue."

69. Lancaster, *Making Time*, 8, 331–33.

70. Martha Moore Trescott argues that American women have not provided the engineering field with significant numbers, but rather significant theoretical contributions as "systems thinkers." See "Women in the Intellectual Development of Engineering," 147–81, 149.

71. The Gilbreth children graduated from the following institutions: Anne, Michigan; Ernestine, Smith; Martha, New Jersey State College for Women; Frank, Michigan; Bill, Purdue; Lillian Jr., Smith; Fred, Brown; Dan, Penn; Jack, Princeton; Bob, North Carolina; and Jane, Michigan.

72. Gilbreth, *As I Remember*, 234; Yost, *Frank and Lillian Gilbreth*, 339–41.

73. Gilbreth and Carey, *Belles on Their Toes*, 80–81.

3 To Embrace or Decline Marriage and Family: Annie Jump Cannon and the Women of the Harvard Observatory, 1880–1940

> If all the ladies should know so much about spectroscopes and cathode rays, who will attend to the buttons and breakfasts?
>
> —Senior European astronomer to Sarah Whiting,
> chair of astronomy at Wellesley, 1880s–1890s[1]

> While we cannot maintain that in everything woman is man's equal, yet in many things her patience, perseverance and method make her his superior. Therefore, let us hope that in astronomy, which now affords a large field for women's work and skill, she may, as has been the case in several other sciences, at last prove herself his equal.
>
> —Williamina Paton Fleming, "A Field for Women's Work in Astronomy,"
> World Columbian Exposition, Chicago, 1893

> We work from morn 'till night
> For computing is our duty
> We're faithful and polite,
> And our record book's a beauty;
> With Brelle and Causs, Chauvinet and Pierce,
> We labor hard all day;
> We add, subtract, multiply, divide,
> We never have time to play.
>
> —From the parody *Observatory Pinafore*,
> written by women computers
> of the Harvard Observatory[2]

IN 1931, A JOURNALIST FOR THE *CAMDEN DAILY COURIER* HERALDED THE recent recipient of the Henry Draper Medal for research in astronomical physics. For the first time in the history of the prize, it was a woman, the ebullient Annie Jump Cannon, curator of photographic plates at the Astronomical Observatory of Harvard University. That a woman would be so honored was astounding. Since most science fields had been professionalized toward the end of the nineteenth century, women had found themselves relegated to teaching posts at women's colleges, or they lingered imperceptibly at the lower echelons of the promotional ladder as lab technicians, or they quietly worked as assistants to husbands in the few home labs that remained. Without PhDs, they could not advance in rank or title; most could not even put their names on papers they researched and wrote themselves.

Annie Cannon was an anomaly to be sure, yet this journalist assured readers that thirty-five years in science had not tarnished her womanliness: she was affable, unassuming, and hospitable to guests in her home and the observatory both, and doted on children. She seemed to know her place in the natural order. She "would have been a first-rate housewife," he assured readers, but instead "took up light housekeeping among the stars":

> "Oh those untidy men folks," we can hear Miss Cannon say as she took up astronomy. "Let's get some order in this kitchen, I mean heaven."
> So she made her life work the cataloguing of the stars. Hundreds of thousands of them she "dusted off," as it were, and put back into their right places. . . .
> Housewives may be a little weak on astronomical physics. But they will understand just how Miss Cannon felt. Those heavens simply HAD to be tidied up.[3]

Later in the century, sociologists would theorize that a child's scientific proclivities come from the empirically minded patriarch of the family, but that was not the case with Annie Cannon. Her mother was the parent who opened the trapdoor to the roof of their Delaware home so that they might gaze at stars. Elizabeth Cannon encouraged her daughter to study mathematics, chemistry, and biology at Wellesley College, and eventually to study physics under MIT-trained astronomer Sarah Whiting. She was pleased when her daughter returned to Wellesley and Radcliffe for postgraduate courses, and ultimately accepted her daughter's choosing vocation over marriage. At thirty-four, Annie Cannon excused herself irrevocably from a traditional life as wife and mother by accepting a post at Harvard's astronomical observatory.[4]

Thirty-five years later the Daily Courier lauded her as a great scientist, rather than mocking her spinster's status. The praise was strange, considering that both men and women thought disinterested objectivity, the hallmark of the modern scientific expert, was achievable by men alone. In Cannon's day, anthropology, botany, and nutritional science seemed appropriate disciplines for women "assistants," since these fields could be viewed as germane to homemaking.[5] So why would anyone extol Cannon's womanly traits in astrophysics—a field so virile that women would be denied access to telescopes in the nation's observatories until the 1960s?

Within this male-dominated field was a domain in which presumably female traits such as patience, a tolerance for tedium, and general domesticity became assets. It was not a prestigious enclave and appeared supportive and peripheral to the work of credentialed scientists. Yet from this enclave Cannon found a way to revolutionize the field. She "laid the foundations of systematic study of the stars for all time to come," the *Daily Courier* exalted, not out of her brazen disregard for women's station, but just the opposite—out of her womanly proclivity to tidiness.

Compared with European astronomy, the American field that Cannon entered in 1896 was unorganized and small scale, not yet the full-blown professional science it would become in the twentieth century; but the Harvard Observatory was about to become the hierarchical bureaucracy that characterized the new era of "big science." Edward Pickering, an MIT physics professor, hastened the transformation when he became director in 1876, as did the advent of stellar photography. Amateur astronomer Henry Draper developed a method for photographing lines of stellar spectra viewed from telescopes onto glass plates, providing permanent, portable images of the sky. Pickering departed from the old astronomy of charting locations; he used the new photographic plates to focus on stars' brightness and color as clues to their distance and chemical composition. Women proved to be some of the most generous financial supporters of his projects. Catherine Bruce donated fifty thousand dollars for the erection of a photographic telescope, and Anna Palmer Draper, widow and former assistant of Henry Draper, donated large sums of money to the observatory for the Draper Memorial, a catalog of stars compiled through the technology pioneered by her husband.[6]

Pickering collected an overwhelming library of photographic plates from outposts in Cambridge and Peru, and he needed processors of the stellar information they contained. These "computers," as he called them, would not be "astronomers." He viewed them as doers, not thinkers; workers, not scholars; amateurs, not professionals with the expectations of compensation and promotion that professionalism brings. They were recorders of measurements taken from photographs, providers of raw data for trained astronomers to interpret. The work was not belit-tling in Pickering's eyes. His Baconian worldview had convinced him that the best science was that garnered through piles of facts, and his workers had the important job of gathering an abundance of data to achieve scientific truth. They would have the basic computational skills of high school and college graduates, rather than formal training in as-tronomy, making them three and four times more affordable than cre-dentialed men. Most important, they would have patience and a pen-chant for detail, traits Pickering believed inherent in the female sex.[7]

Pickering's instincts about women were not novel at the time. Earlier in the nineteenth century, the study of stars, like the study of flowers or insects, seemed an appropriate avocation for women already deemed close to nature. Fathers and husbands observed alongside women from telescopes in attics or on city walls; the respective labor of men and women had not yet been perceived as divided between thinking and doing, analyzing and observing, professional and amateur. George Greenstein has estimated that between 1859 and 1940 one out of every three American astronomers was a woman, yet only a handful has been acknowledged in the historical record. In 1847 young Maria Mitchell was studying the heavens from a telescope at her Nantucket home when she spotted the comet that made her famous. She received a gold medal from the king of Denmark, recognition that propelled her in ways fu-ture amateurs would never know. Among other honors, in 1848 she be-came the first woman member of the Academy of Arts and Sciences, and in 1850, of the Association for the Advancement of Science. In 1865, she became the first woman professor of astronomy at Vassar College. Mitchell made an independent income computing tables of the posi-tions of planets for the Coast Survey, and she did it all from home.[8]

Two of Mitchell's most promising students, Mary Whitney and Antonia Maury, left Vassar for Harvard, the former to study and the latter to innovate the classification of stellar spectra. A handful of other protégées were permitted to enter the Harvard Observatory to compute for the Durchmusterung catalog, and by 1886 an entire female com-puting staff worked on classifications for the Draper Memorial. The abundance of plates to be processed forced Pickering to become a good industrialist, eager "to secure the greatest possible output for every dol-lar expended." The prevailing assumption (though often not the real-ity) was that women who worked outside the home were earning wages supplemental to those of male primary breadwinners. Pickering paid his hires accordingly: twenty-five cents an hour for seven hours a day, six days a week. Although a respectable wage for factory workers, it was lackluster for retail, secretarial work, and other pink-collar trades. A few especially skilled women received a salary of six hundred dollars a year, two hundred dollars less than male assistants who engaged in "mechanical" work. But for the most part, twenty-five cents an hour remained the going rate until 1906.[9]

Exploitive as it seems, many women viewed computing as an opportunity too good to refuse, and Pickering never lacked for bright applicants. His most talented computers, first- and second-generation graduates from elite women's colleges, found few professional alternatives. A handful of their colleagues filled positions at women's colleges, but computing allowed a devoted scientist to do rather than teach. Like no other American observatory of the early twentieth century, Harvard had become the inner sanctum of astronomical advancements, and the busy work of women computers was largely the reason, since their data provided new answers to new questions. By 1893, seventeen assistants at the observatory were women, twelve of them responsible for identifying spectral phenomena in photographs, computing the measurements, and preparing results for publication. One hundred and sixty-four women worked as computers in American observatories between 1875 and 1920. The Dudley Observatory in Albany, the Yerkes Observatory in Washington, D.C., as well as the Lick, Columbia, Allegheny, and Yale Observatories opened positions to women, no doubt because of the success of Pickering's experiment. By 1921 eight of the twenty women astronomers listed in *American Men of Science* were computers as Pickering defined them. Women had found their niche in this masculine field of physical science.[10]

Annie Cannon was sure that the development of stellar photometry was what made her scientific career. She could obtain access to the dry-plate photograph's record of the sky regardless of her proximity to telescopes and university classrooms. Stellar photometry turned astronomy into a "daylight profession," she explained, "more suitable for women than when eye-observations had to be made at night in many sorts of wind and weather." Indeed the technology allowed "women to remain women"; they need not brave the harsh elements to study the stars, and they could engage in a brand of detailed busywork befitting of mothers and wives who would otherwise be measuring ingredients for a cake or counting needlepoint stitches. Scientific photography had created specialties for women in other fields too. Cannon's contemporaries Katherine Foot and Ella Strobell, for example, began the innovative use of microphotography in biology when hand drawings were still the practice.[11] The inherent tension provoked by their leaving home for professional science was eased, since as observers of photographs, the women appeared to be doing the work of helpmeets rather than being interpreters or experts in their own right.

Before Cannon arrived at Harvard, Williamina Paton Fleming had looked at photographic plates and classified spectra for the *Harvard Annals* and observatory circulars. Cannon built on Fleming's observations, sharpened her classification schemes, and codified twenty-eight spectral classes and subdivisions for the four hundred thousand stars of the Draper Catalogue and Draper Extensions. Cannon's work had to be conceptual—for how else could she have named her spectral categories—but Pickering continued to describe it as a form of observation that required little specialized understanding of stars. The women computers didn't argue, though they thought that Cannon's special talent could not be matched by men. The plates from which Cannon had worked provided little more than random dots, smears, blips, and shades of gray, yet with magnifying glass in hand she rapidly labeled the marks and called out identifications to women who frantically kept ledgers. Margaret Mayall, a chief assistant, thought Cannon's visual memory innate, not learned. "She had wonderful eyes," she recalled, "and she could see things that very few people would recognize until she pointed it out." Another colleague likened Cannon's visual recall to having a phenomenal memory for faces: it was not based on reasoning; "she simply recognized them." It was sensitivity for nuance that men just didn't have.[12]

Male astronomers previously had based their classification systems on theories about stars' life cycles, but the development of photometry made it possible for Cannon to discern differences that corresponded to the temperature of stars, their chemical composition, and other physical features. Her categories, which were formally adopted as the standard at the 1910 meeting of the International Astronomical Union, weren't named after her, as men's innovations are often named after them; hers was known as the "Harvard System."[13] The mnemonic "**Oh Be A Fine Girl, Kiss Me**" reminded astronomers of the labels for her categories, but also of her place in the social order of the observatory. Cannon accepted her label as "a fine girl" among males of supposedly superior intellect. Her matronly charm made her appear the embodiment of domesticity. She mothered the younger women and hosted visiting scholars of the observatory as if they were part of her extended family. She classified spectral phenomena on plates with the precision of a needleworker, presumably without interpreting what she saw to any scientific degree. It was not unusual for her to leave the observatory at six o'clock and return as others turned in for the night. She broke down boundaries between home and workplace, science and domesticity by collapsing the distance and time separating them to almost nothing.

Cannon was both an innovator and hostess at one of the world's leading astronomical centers; she brokered partnerships and exchanges of equipment between men in the international community and assumed an ambassador-like role outside it. Male astronomers had her to thank for helping to organize international meetings, but also for increasing their visibility and status in the popular media. She prepared scripts for radio about the study of stars and wrote books and articles for laypeople and children. It is not a requirement that science popularizers be women, though historically this has often been the case. In Mitchell's day English women such as Agnes Clerke popularized the study of the skies, and Margaret Huggins, wife and assistant to astronomer Sir William Huggins, called for a new class of "science worker," one who served as historian, critic, expert, and originator of scientific ideas all combined. Cannon also embraced her cultural work as an educator. As a woman on the professional periphery yet a major figure in the vanguard of discovery, she enjoyed a dual perspective men didn't share. You live in "the meridian of things astronomical," a colleague remarked, "ever alert to keep up with surprises that come in by letter and cablegram from everywhere—comets, novae, et. cet . . . [yet] it is gratifying to see that popular magazines and the Sunday editions are alive to the growing interest of the general public in star science."[14]

Cannon was supposedly an organizer, not a conceptualizer; a translator, not a creator. And yet her work was so significant that it made her famous. In Europe she became a member of the Royal Astronomical Society and received honorary degrees unprecedented for a woman of any nationality. Back at home Wellesley College admitted her as an honorary member of Phi Beta Kappa, and her portrait occupied "the place of honor" on the front pages of *Scientific American* in 1929. Two years later she was voted one of America's twelve greatest women, and two years after that she represented professional women at the World's Fair in Chicago. With Amelia Earhart, Mary Beard, and Jane Addams, she spoke on "the contribution of women to civilization," her part devoted to women in the natural sciences.[15]

Preparing for this appearance, she wrote the chairs of science departments of women's colleges, asking for lists of the most accomplished women scientists of the past one hundred years. The responses were discouraging. The director of the Smith Observatory could think of pedagogues, but no researchers. Mount Holyoke women came up with one of their own, zoologist Cornelia Clapp, but that was it. The faculty of the Wellesley zoology department came up with Florence Sabin but lamented that much of her work had been attributed to others. Colleagues in the chemistry and physics departments both claimed Curie as their sole female standout aside from Ellen Swallow Richards and the German physicist Lise Meitner. ". . . We have no Madame Curie in botany" wrote the Wellesley department chair. The Curie complex continued.[16]

Cannon was hard pressed to tell of women's greatness in physics or chemistry at the World's Fair, but she had plenty to say about women in astronomy. "There is nowadays hardly an Observatory, especially where research is undertaken, without one or more women on the staff," she reported. To her count, women had discovered forty-two hundred variable stars; of new stars, thirty-five had been discovered by Harvard women compared with sixteen elsewhere. Women studied at the observatories of Princeton, Michigan, Virginia, and Yale, and she avowed that the feminization of astronomy was not merely a national trend. The chief of the photometry bureau at the Sorbonne was Dorothea Klumpke, a woman who had taken her doctoral degree there in 1893, several years before Curie took hers in physics.[17]

While Cannon's significant work in her field could be viewed as a positive signal for women, one could also read the achievement differently. The field of astronomy had opened doors for women, but only to allow them to work in limited capacities. In the eyes of peers, computers were not transgressing but rather transferring domesticity to the observatory so that men might freely conduct the interpretive work of real science. Cannon embraced the supportive role offered to her. "Whose eye is keener than a woman's to examine these photographs?" she asked. "Whose hand is more deft to handle this precious and unique library of first and only editions?" And yet for all the accolades she attributed to women at the World's Fair, she conceded that men "held the lead" in "interpretive" enterprises.[18]

This makes Annie Cannon a complicated figure—a feminist and traditionalist in one. It became her special cause to make it possible for more women to live out their dreams of science. She traveled across the country, speaking to women's groups, asking for no compensation beyond train fare, hoping to reach amateur stargazers waiting to break into the field. As a charter member of the Maria Mitchell Association, she doled out scholarships to women such as A. Grace Cook, an enthusiast who spotted meteors through an outdated telescope from a deck chair eighteen miles from home. When Cannon won the Ellen Richards Prize of the Association to Aid Scientific Research by Women in 1932, she gave the prize money to the American Astronomical Society to support women astronomers.[19] And yet, for all her achievements, Cannon never challenged women's roles in science once they had gained entry to its institutions.

Williamina Fleming, Antonia Maury, and Henrietta Leavitt: The Politics of Invisibility

In demurring to men Cannon had won universal respect, but some of her female colleagues were not demurring types. Williamina Fleming wanted the pay of men; Antonia Maury wanted their prestige; and Henrietta Leavitt wanted men's God-given right to theorize about the natural world for herself. As inappropriate women, their careers have been far more obscured in the historical record than Annie Cannon's. Their stories tell cautionary tales, the flip side of being a woman pioneer in masculine science.

Without academic degrees, Williamina Fleming taught Cannon all she knew when Cannon first arrived at Harvard at the turn of the century. Fleming was a Scottish immigrant who came to the United States with her husband in 1878; within a year, she was abandoned, pregnant, and destitute. To survive, she found domestic work, fortuitously for Edward Pickering. Her timing was also fortuitous, for he was increasingly dissatisfied with his male assistants and impulsively declared one day that his twenty-four-year-old housekeeper could do their job better if given the chance. To prove his point he hired Fleming at the observatory in 1881, and she was grateful to work as his experiment. She started by copying records and doing the most elementary of computations, and within five years she could replace head computer Nettie Ferrar, who left the observatory to marry. In honor of the man who saved her, Fleming gave her son Pickering's name and encouraged him to study science; he went on to study engineering at MIT the same year his mother was promoted to curator of the Draper Catalogue, the first formal appointment of a woman in the history of Harvard. His mother had twenty women in her charge.[20]

What led Pickering to try such a risky experiment? Apparently he thought Fleming's domestic skills would be functional in his observatory. Known to craft dolls in intricate Scottish Highland costumes, this meticulous needleworker brought order to Pickering's plates just as she had brought order to his home. In addition to filing the plates, she examined each with a magnifying glass, noting peculiarities in the spectra that male staff members had never attempted to find. Much like organizing buttons by shape and color, she devised twenty-two classifications for more than ten thousand stars, making it that much easier for others to study them. In the sorting she discovered three hundred variable stars, ten novas, and fifty-nine gaseous nebulae, all of which made their way into the Draper Catalog.[21]

Fleming had had no formal training, yet a contributor to the *Chicago Examiner* called her an American Madame Curie. Both women had proved that "when a woman does turn her mind to science she brings into the field qualities that her proud 'lord and master' cannot match." Professional astronomers agreed. Fleming went on to become an honorary member of the Royal Astronomical Society and an honorary fellow of Wellesley College. Her work was painstaking, yet the journalist for the *Examiner* described it as perfectly suited to her female mind: "The amount of labor devoted by Mrs. Fleming to the patient study, under the microscope, of thousands upon thousands of photographs of the spectra of stars would have daunted most men and would have been utterly beyond their capacity, for it require[s] a combination of patience and persistence, and faith, and minute accuracy which is, perhaps, rather a feminine than a masculine characteristic."[22]

The ability to multitask was apparently yet another of her womanly characteristics. Not only did she perform the work of photometry, but she also edited publications and acted as Pickering's personal secretary. She chronicled her rigorous daily routine on a March day in 1900:

> Observatory from 9h 05m to 4h 45m. Final notes on Vol. XLV [of Annals], 0h to 6h relating to variable stars. . . . The work on suspected variables involved their examination on a number of photographs for confirmation of variation, and this work I continued with Miss Leland[,] some having already been done. Then I did some work with Miss Woods on "out of focus" plates, and later with Miss M. Stevens on the southern meridian photometry. Then went over the latter work with the Director, adopting constants, and entering and discussing atmospheric absorptions. We found that in the earlier part of the year, when two series of observations were taken on the same date (one early in the evening[,] the other in the morning) the value of the constant was smaller in the evening than in the morning, while towards the last part of the year the value of the constant was greater in the evening than in the morning thus seeming to indicate some change due to the season of the year. . . . We were able to read through a few more pages of Miss Cannon's "remarks" before the Director left at three o'clock. I then gave some time to Mr. Bailey on the work on the magnitude of his variables. Then returned to Miss Woods and "out of focus" plates.[23]

This entry reveals a relationship between Fleming and male astronomers more complex than that of a subordinate to her superiors. While men ultimately published her data under their names, interpretation of the data was increasingly a collaborative endeavor.

Men rewarded her with praise, but there were clear limits to how far her womanly assets could take her. In 1900 her salary was fifteen hundred dollars, generous compared with the compensation of the other women computers but insulting compared with the twenty-five hundred dollars a year male assistants received. Her salary was based on a seven-hour day, but she typically worked nine hours or more, while men left on time. She observed that many of the men "took things easy" at work. They operated at a slower pace by day and went home to wives who catered to them at night. "My home life is necessarily different from that of other officers of the University since all housekeeping cares rest on me," she pointed out. She employed a maid to lighten the load, but given her lesser pay, she couldn't afford one full time. On Sunday, her day of respite, she madly prepared food for the week and did the washing, for she hadn't a spouse to do it. Sometimes she couldn't get it all done and would have to leave the observatory midweek to pay bills and buy groceries for supper.[24]

By 1900 her financial situation had become untenable. Her son, Edward, still had another year at MIT; she couldn't maintain her home and pay his tuition without a raise. Pickering agreed that Fleming was a rare breed of workhorse, who took few vacations or holidays, but her salary had nearly doubled since her promotion to curator and she hadn't the credentials to earn more. "He seems to think that no work is too much or too hard for me no matter what the responsibility or how long the hours," she seethed. "But let me raise the question of salary and I am immediately told that I receive an excellent salary as women's salaries stand." She might have been content to proceed at her level of pay had she been able to pursue variable stars, but she was relegated to other tasks. "Looking after the numerous pieces of routine work which have to be kept progressing, searching for confirmation of objects discovered elsewhere, attending to scientific correspondence, getting material in form for publication, etc., has consumed so much of my time during the past few years that little is left for the particular investigations in which I am especially interested," she carped. "The Director, however, says that my time employed in the above work is of more value to the Observatory."[25]

Fleming thought she would try a temporary solution: she'd do routine secretarial work during the day, and she'd take the variable-star work home at night. In 1903, after regular working hours, she classified and took light measurements for more than thirty-five hundred stars. Still, her plan failed to bring her an increase in salary. In fact, by doing observatory work at home, she had helped to render it invisible; her work lost its value as scientific endeavor, as "specialty," as "work" at all. The long hours she labored wore her down physically and emotionally during her final decade at the observatory. Friends urged her to reconsider how much time she devoted to Pickering, but she would never disappoint him. Nurses had to exercise an arm she could barely move, so that she could complete assigned work on the plates. Ceaselessly, she continued until 1911, when, by all accounts, she had worked herself to death.[26]

Her martyrdom was not lost on newspapermen. Work like hers was never glamorous, one wrote—"no glad cry of discovery as some wonderful new star bursts upon sight. . . . Her work was simply poring over faint and ill-defined photographs, peering, and noting, and comparing, and adjusting the light for better seeing, until the overtaxed eyes could do no more." While male astronomers attended conferences and filled their days with a varied diet of activities, Fleming did the same tasks over and over again. Only a woman could show this degree of devotion to science, this journalist believed, and do it without the incentives of money, fame, or discovery that motivated men. Pickering, too, extolled her virtues in eulogy. Along with raising a son, she had cared for her mother, and her brother's children, kept a spotless home, and "had a re-markable skill and artistic taste with her needle." But the domesticity he attributed to her was a double-edged sword. It was both the root of her competence as a computer and the assigned cause of her inadequacy as a scientist. It's what gave Pickering license to publish all but two of her papers in his name. She "formed a striking example of a woman who at-tained success in the higher paths of science without in any way losing the gifts of charm so characteristic of her sex," he wrote. He could pay no higher compliment for her years of devoted service.[27]

For a time Williamina Fleming and Annie Cannon had a collaborator named Antonia Maury, a woman who sought to advance as male astron-omers did. The niece of Henry Draper and a student of Maria Mitchell, she thought herself well positioned to be the exceptional computer-turned-legitimate-scientist. After graduating from Vassar in 1887, she went to Harvard and started classifying the spectra of bright northern stars as part of the Draper Memorial. Almost immediately she made her ambitions known and proved eager to publish original research rather than to classify for classification's sake. She abandoned the spectral cat-egories Pickering had assigned her for a two-dimensional system that accounted for factors of width and sharpness of spectral lines. Ever the scientific positivist, Pickering thought her system "unnecessarily elabo-rate," prone to subjective shades of gray when Annie Cannon's single dimensions presented distinctions in black or white. He was annoyed that Maury would question a good thing, but also that she would try to impose her own way of seeing his plates so that she could be recognized as a scientist in her own right.[28]

Eventually astrophysicists used her scheme to recognize subtleties in size and luminosity that would have been indistinguishable under earlier systems of classification, but recognition at Harvard eluded her. When her experience made visible the glass ceiling for women at the observatory, Maury's reaction was to bristle, for she felt entitled to better treatment. She told Pickering that her situation was unacceptable, and that she was willing to resign, but that she would not give up her projects until he formally credited her ideas: "I should have full credit for my theory of the relations of the star spectra and also for my theories in regard to B. Lyrae. Would it not be fair that I should, at whatever time the results are published, receive credit for whatever I leave in writing in regard to these further matters?" Shocked by her audacity, Pickering agreed to give her time to complete her original research rather than to pass it on. Illness delayed her and confrontations continued—but a squeaky wheel does eventually get its grease. In 1897, Antonia Maury was the first woman to receive formal recognition for her investigations on the title page of the observatory *Annals*.[29]

Although Maury won her battle, she didn't win the war, for she, not Pickering, capitulated about the best method for classifying stars: "I, being naturally unsystematic[,] was not able to understand what you wanted," she told him. Still, she couldn't help but qualify her concession: "You also, not having examined minutely into all the details, did not see that the natural relations I was in search of could not easily be arrived at by any cast iron system."[30] Maury had pored over the plates and discerned minute differences in the spectral lines of stars, much as the geneticist Barbara McClintock discerned the minute details in kernels of corn decades later. Both women would know the frustration of men's unwillingness to see nuance in the natural phenomena they studied. They were women who observed intimately. One's knowledge of corn mutation, like the other's knowledge of spectral lines, was almost intuitive, and male scientists didn't trust their intuition.

With nowhere to go, Maury begged to return to the observatory part time; Pickering took her in but complained about her perpetual tardiness thereafter. Her work on the Draper Memorial was belabored because of frequent illness and worsening eyesight, her progress on the northern stars not nearly as rapid as Annie Cannon's on the southern ones. Maury's instincts about spectra were just as keen, but she slowed down the process of classification even more by asking what it all meant. Cannon never asked, and as a result may have achieved more successfully than anyone the masculine disinterestedness that Pickering revered. She had no preconceived theories to which she hoped her data would conform. She identified, measured, recorded, and moved on; witnesses called her process virtually instantaneous. Although she was a warm and gregarious woman, she became hermetic around the plates, removing her hearing aids to avoid human distraction or influence. She worked too quickly to leave traces of human agenda or attachment behind. Maury, meanwhile, couldn't help but wear her curiosities and dashed hopes on her sleeve.[31]

Anna Draper did not defend her niece, but rather apologized for Maury's unacceptable attitude, for clearly Maury had broken the silent compact keeping womanly ambition in check. In the decades that followed, Maury was a presence in the observatory, but she was perpetually underappreciated and misunderstood. The astronomer Cecilia Payne-Gaposchkin described her in later years as "a dreamer and a poet," unconcerned with whether her socks matched but anxious to speak her truths to anyone who'd listen. If only Pickering had let her learn algebra or publish more papers or perform more original research, she'd pine. After his death the International Astronomical Union added the letter c spectral type to the others, delineating the differences in spectral lines that she had called for thirty years before. The American Astronomical Society awarded her the Annie Jump Cannon Prize for her research when she was seventy-seven years old. These small nods of recognition sustained her until she died in 1952.[32]

Henrietta Leavitt, a woman nearly the same age and temperament as Annie Cannon, was also deaf and worked without the distraction of noise or Maury's ambition. Like Cannon she also lived without children or husband and within walking distance of the observatory, giving her that much more focus on the business of classifying stars. She had attended Oberlin and Radcliffe Colleges and was drawn to astronomy by course work. The twenty-five-year-old asked Pickering if she could volunteer at the observatory in 1893. She was of independent means, she assured him; all she wanted was to get closer to the spectral plates. Stellar photometry suited her from the start; a Princeton man thought her a "variable star fiend," the way she sat hour after hour poring over the plates. She proved her worth; Pickering rewarded her with a paid position in 1902 at five cents better than the going rate.[33]

The Magellanic Clouds soon became her favored corner of the universe, and she discovered 1,777 stars in the region. As more photographs arrived from the observation outpost in Peru, she began to notice a striking correlation between the periods and luminosities of the stars captured in them. Cannon was less mindful of the correlation, but Leavitt saw that having a known value for a star's rhythm allowed her to calculate its inherent magnitude. If she compared this value with the known distances of closer stars, any star's distance could be calculated. The implications were enormous, but Leavitt was paid to measure and record, nothing more. In a brief paper she furnished for the Harvard *Annals* in 1908 she made it clear that she understood her place. Only in passing did she suggest that the longer periods of the brighter variables were "worthy of notice"; astronomers would "probably" observe the "apparent" relation between periods and emission of light if they cared to explore them.

Her passive language didn't fool astronomers who read the report; letters came in from all over the world asking Pickering for more data. It wasn't forthcoming, for he had reassigned Leavitt to the classification of stars in less exciting parts of the galaxy. Moreover, her failing health caused her to seek rehabilitation at a family farm in Wisconsin. Pickering was desperate for her to process photographic plates of the North Polar Sequence and posted them to her, and she garnered the strength to work on them two to three hours a day. The pace was sufficient to publish results under his name in the *Harvard Observatory Circular* in March 1912. Like Annie Cannon, Leavitt did what she was told and never complained.[34]

It took her years rather than months to provide the measurements verifying the period-luminosity relationship she theorized in 1908. Men, meanwhile, swooped in to exploit her initial observation. Young Princeton astronomer Harlow Shapley used her theory to make measurements in the Milky Way and ultimately to determine the size of the galaxy, an accomplishment that propelled him to the forefront of the field. Replacing Pickering as director of the Harvard Observatory in 1921, he expressed gratitude for Leavitt's pioneering work, even as he passed much of it off as his own. She observed, he maintained, but left him to think through the implications. In 1925 a mathematician from Sweden wrote to tell Shapley that he was seriously inclined to nominate Leavitt for the Nobel Prize. News of her idea about the period-luminosity relationship had reverberated throughout the international community, but apparently news of her death had not; Leavitt died four years before Shapley received the letter. It's unlikely she ever knew the full impact of what she discovered.[35]

Cecilia Payne-Gaposchkin: New Woman, New Astronomer, New Politics in the Stars

Annie Cannon, Antonia Maury, and Henrietta Leavitt were first-generation college-educated women; their place as pioneers was all too apparent in their limited choices after college. For them, career and marriage were mutually exclusive paths, and the first computers of the Harvard Observatory had decidedly chosen the former over the latter or they had left work when they married. Male scientists of their generation, like Pickering and his male assistants, were married and had children, along with better salaries and promotional prospects. For later generations, the tensions between computers' identities as women and professionals, as domestics and scientists, eased, but were never fully reconciled. More women combined marriage and career, or worked until children competed for their time. More of them entered coeducational doctorate programs and expected research careers afterward; they wanted to compete with men in theoretical science, to place their names on original research, and to attain equivalent promotions. They asked to be acknowledged as scientists, and sometimes they weren't disappointed.

The percentage of married women listed in *American Men of Science* rose in the 1920s and 1930s, though unemployment rates for married women rose as well because of nepotism policies that prevented wives from working in the same academic departments as their husbands. At the Harvard Observatory, an increasing number of women graduate students married men of similar rank, and in the 1950s, one of them, Cecilia Payne-Gaposchkin, became the first female full professor appointed at Harvard. Born in 1900, she was a generation or two younger than Annie Cannon and represented a new breed of professional woman scientist. Rather than perpetuate a separate female sphere at the observatory, she sought to become integrated with male astronomers in ways her predecessors never could. Annie Cannon had been childless but outwardly maternal; Cecilia Payne-Gaposchkin bore three children but appeared mannish by comparison. She stood on a robust five-foot-ten-inch frame, kept her hair short, and did not accentuate her womanly attributes. Cannon was bubbly and warm; Payne rarely laughed and avoided emotionally charged subjects. It may have been her British reserve, but it was likely also her way of attempting assimilation into the professional world of astrophysics.[36] While her career testifies to the ability of women to succeed on the same terms as men, it also shows the vulnerability of women attempting integration. From a man's perspective, the process could never be complete.

Growing up in rural England, Cecilia Payne knew she wanted to be a scientist. Her mother, like Annie Cannon's, had cultivated her daughter's love of nature. Emma Payne introduced her daughter to spiders, mimosas, and orchids; one day with her daughter still in the pram, she pointed to a meteor blazing across the sky. Unfortunately, schoolteachers attempted to dissuade the child from pursuing science, since they viewed it as antithetical to faith in God. One nun thought the girl would be "prostituting her talents," but stopped discouraging her once Payne developed facial hair: in lieu of marriage prospects, thought the nun, science might become the young woman's best means of support. "You've got brains. Make something of them," the family doctor told her. It stiffened her resolve to become a scientist.[37]

In grammar school Payne was drawn to botany, but at Newnham, the women's college of Cambridge, she moved into physics. Under Ernest Rutherford, the Cavendish laboratory attracted a well of international students. While many subdisciplines of physics were developed further in Europe than in the United States, tolerance for women physicists was nascent at best. Payne's experience in Rutherford's lecture courses was similar to Marie Curie's twenty years earlier at the Sorbonne. Payne sat separately in the front row of the lecture hall, ever reminded of her trespass. In labs she paired off with other women and felt like a perpetual "second-class student." Since pure physics seemed too hostile, she decided to become an astronomer.[38]

Payne biked up to the observatory and declared her intentions to Arthur Eddington, "the father of modern astrophysics." Amused and skeptical, he still agreed to help her prepare. Another rare, sympathetic advisor feared she would never be taken seriously by English academics. Most women in her class took up teaching positions after graduation, "a fate worse than death," as she saw it. She looked to the United States, where she would pursue a research career, leaving Cambridge, England, for Cambridge, Massachusetts in 1923. Harvard Observatory director Harlow Shapley was intrigued by assurances from her advisor that, "given the opportunity, [she] would devote her whole life to astronomy. . . . She would not want to run away after a few years training to get married." Shapley offered her a fellowship, figuring that when it ended, he would have another Annie Cannon at his disposal.[39]

When Payne arrived at the observatory, the ghosts of Pickering's original "harem" still lingered everywhere she turned. A few of the old "strutting hens" remained: Miss Woods, Pickering's former secretary, still displayed the medals for the novas she had discovered; younger women told Payne that the spirit of Henrietta Leavitt haunted the stacks. Payne took Leavitt's old desk, and hoped not to share her fate; she thought Pickering had condemned a "brilliant scientist . . . to uncongenial work" and that his decision to hire her to "work, not think . . . set back the study of variable stars for several decades." This wouldn't do for a self-professed "theorizing" type. A friend from Newnham College described Payne as perpetually absorbed in abstract questions. Sometimes she would sit in bed playing the fiddle or constructing mechanical toys as she worked through complex problems in her mind. She wanted to grapple, to explore, to be "in direct touch with the fountainhead."[40] Under Shapley it would be difficult. Since he had taken over the directorship the female computing corps had nearly doubled in size; twenty-five assistants worked full time, and he had exploited Radcliffe undergraduates during the school year, since "girl hours" were less of a blow to the budget. Payne worked as an independent fellow in her first year at Harvard, thus free to study the temperature and composition of stellar atmospheres as she wanted. But once the fellowship ended, she was at Shapley's mercy, and it was soon clear that she'd measure stellar magnitudes all day like the rest of the women computers.[41]

Payne feared that falling into standard photometry would make her indistinguishable from everyone else, but Shapley saw that she had special talent and encouraged her to pursue a doctorate degree. There was not yet an astronomy program at Radcliffe or Harvard, so she turned to the physics department and once again felt the hostility of men who thought she was trespassing. The department chair wouldn't admit her, so Shapley brought together a makeshift committee of doctoral examiners and shaped a graduate program in astronomy. Other women with whom she studied stopped at the master's degree and married; Payne, however, became the first recipient of an astronomy PhD at the college. She, like Annie Cannon before her, had become a pioneer of sorts, and she shared the ambiguous status that comes with being first.[42]

Like no other member of the observatory staff, she straddled worlds between photometry and spectroscopy, female doing and male thinking, tedious classification and inspired theory. The result was a dissertation in 1925 that married Pickering's pile of facts and Shapley's penchant for synthesis. She applied the ionization theory of physicist Meghnad Saha to the range of spectra that women computers had identified over the years but left unanalyzed. She concluded that stellar atmospheres were not made up of metals, as had been believed; rather they consisted primarily of hydrogen, with a bit of helium and traces of other elements.

When she showed a preview of her write-up to astrophysicists Arthur Eddington and Henry Norris Russell, they advised her to temper her stance. "It is clearly impossible that hydrogen should be a million times more abundant than the metals," Russell challenged. His stature in the field made Payne question herself, lose heart, and renege; in the final dissertation she made drastic revisions and decided that the stellar abundance she deduced for hydrogen and helium was "improbably high . . . almost certainly not real." Four years later, when Russell's data were in "gratifying agreement" with Payne's original contention, he was gentlemanly enough to cite her work, but the credit for the discovery was his. A colleague offered years later to set the record straight in a history of astrophysics, but she humbly refused. The mistake was hers for not pressing the point, she told him. Her self-doubt speaks to the age and gender politics involved.[43]

Until now gender and rank at the observatory had been virtually synonymous, but Payne's accomplishments in theoretical astrophysics muddied the waters. She was earning an international reputation; men thought her dissertation a sure classic in the field. She had become the youngest astronomer starred in *American Men of Science*, and in 1928 she was elected to the International Astronomical Union's Committee on Spectra. Still, university president A. L. Lowell swore that she, or any woman, would never get a university appointment while he lived to witness it. Payne was at once resentful, embarrassed, and resigned to her fate. Given her doctorate, her compensation exceeded that of computers, but remained far less than that of male doctorate holders. To family back in England she reported that she was a self-sustaining scientist, hiding from them that she was pawning off jewelry to make ends meet.[44]

She plodded on with the "spade work" assigned by Shapley. Her next work, a book titled *The Stars of High Luminosity*, she wrote grudgingly; it was not the work she wanted to do. In December 1930 she wrote Russell out of frustration; like Cannon she commanded respect outside the walls of the Harvard Observatory, so why didn't this translate to more credit, compensation, or autonomy inside it?

> I have during the past four years had a very unhappy time at Harvard; the chief reasons have been (a) personal difficulties within the Observatory particularly with Dr Shapley, and usually arising out of personal jealousies because he seemed to like others more than myself. (b) disappointment because I received absolutely no recognition, either official or private, from Harvard University or Radcliffe College; I cannot appear in the catalogues; I do give lectures, but they are not announced in the catalogue, and I am paid for (I believe) as "equipment"; certainly I have no official position such as instructor . . . (c) I do not seem to myself to be paid very much; quite honestly I think I am worth more than 2300 dollars to the Observatory. (d) In the seven years I have spent at Harvard I have not got to know any University person through my work (which confines my acquaintance to the Observatory staff and Professor Saunders); whereas the wife of any Harvard man of my status is called upon by the wives of dozens of others.[45]

Shapley tried to appease her with a slight raise in 1931. At twenty-seven hundred dollars a year, she was earning more than twice the salary of women computers, though not quite what a new male lecturer would earn. She demanded time away from Harvard to convene with "intellectual equals" and was granted leave to visit the great observatories out west and to go to see others in Europe. She hoped to clear her head and to get over still another affair of unrequited love. There had been several such disappointments since she first arrived in the States nearly a decade before.[46]

At the Astronomische Gesellschaft in Göttingen she met a Russian-born astronomer named Sergei Gaposchkin. As she spoke to him she was heartened to hear that her scholarly reputation had preceded her, for the man had recognized her name and expressed surprise that she was such a young woman. She thought him physically slight but charming, especially when he told her about his struggle to find a place in Germany where he could do research without being killed. She was so moved by Gaposchkin's story that she vowed to help him leave the country and begged Shapley to hire him. He agreed to create a temporary position, but told Payne that she would have to find the means to keep Gaposchkin at Harvard once the terms of his employment expired.

Payne had developed crushes on male astronomers before, but nothing had ever come of them. Big boned and awkwardly tall, the self-proclaimed ugly duckling had resigned herself to being alone. Her shyness, her chain-smoking, her lack of personal style, however, seemed not to diminish Gaposchkin's awe for her talents. She had a steady salary; an established reputation; a research agenda; and, since 1931, American citizenship to protect him. Colleagues worried that he was taking advantage of a woman clearly desperate for male affection. Within months of meeting, the couple eloped.[47]

It was likely no coincidence that many of the inroads Payne had made in spectroscopy from this moment forward were cut short. When Shapley assigned her husband to photometry, suddenly she was content to do that too, in order to support their joint research. Gaposchkin provided the data for eclipsing stars for their jointly written *Variable Stars* (1938); Payne provided everything else. She claimed that the ingenuity of the work was his, and colleagues thought that was probably the truth, for *Variable Stars* was her least successful work thus far.[48]

In the eyes of colleagues, it was unfortunate that this new breed of woman astronomer had a liability for a husband. Payne's professional admirers understood that she and Gaposchkin were a team and that he was very likely the reason she was denied attractive positions over the years, including the presidency of Bryn Mawr College. Scientists who knew both Payne and her contemporary Irene Curie saw them both in marriages to charismatic men who had pursued—in order to exploit— painfully shy and unattractive women with brilliant minds and professional connections. Frédéric Joliot benefited from the Curie name (he took it for himself) without any added domestic or professional burdens, since Irene Curie took on the child-care duties and refused to give up her lab work.[49] His wife was his link to international physics, just as Payne became Gaposchkin's connection to American astronomy. She wrote up her husband's work in English and shielded him from professional detractors. She never entertained the idea that her marriage was one of convenience, even though most people thought of them as incompatible personalities. "Like two magnets of opposite poles they stuck together," their daughter explained; somehow their differences became complements. Payne found happiness and, for the first time in her life, enjoyed the security of feeling like a "member of a team."[50]

Indeed she had a more collaborative partnership than most; Gaposchkin helped more at home than other men by doling out spankings and doing the lion's share of the housework. He eased his wife's domestic burdens although he didn't remove them fully, as the wives of Harvard astronomers did for their husbands. Payne took on traditional duties when she could; her daughter remembered her as a "domestic wife"—"an inspired cook, a marvelous seamstress, an inventive knitter," who loved to entertain on the rare occasion she was home. Speaking engagements, meetings of the major science societies, and visits to the Yerkes Observatory took her away from the children often, as did the extracurricular activities she refused to give up. When the children were still young she taught Sunday school and played the violin in the Observatory Philharmonic Orchestra. Gaposchkin's salary was always smaller than hers, and hers was of course more modest than her male peers', thus leaving the couple, for most of their married life, in a state of perpetual thrift. They took in borders to ease their burdens and hired domestic workers when they could afford them. Payne resorted to making undergarments out of nylon parachutes she purchased at the army surplus store and became a master canner in the winter months.[51]

Payne was the breadwinner and senior scientist, Gaposchkin the junior partner and helpmeet. The strange arrangement worked but inevitably created its tensions. "There were professional disagreements, not so much on scientific issues as on who should go where when," their daughter Katherine recalled. "She was more important scientifically, so she generally left him as babysitter if there were no other alternatives." At times the situation seemed to insult Gaposchkin's manhood. Male superiors at the observatory thought he had a chip on his shoulder. Toward women assistants he was overly solicitous and fancied himself an athlete with the muscular physique to prove it; they figured this was his pathetic way of compensating. Payne, ever the stoic, seemed to be oblivious to it all; if she thought her husband vain or arrogant, she wouldn't admit it. He needed to be appeased for the whole system to work.[52]

As much as she tried to be a traditional mother and homemaker, the fact was that Payne worked long hours when the expectation was that proper women stayed home with their children. Men at the observatory never questioned her competence as an astrophysicist, but their disapproval of her personal choices as mother and wife was palpable. Shapley was appalled when she scheduled a talk at Brown University in 1940: he thought her simply too pregnant to be discreet. She said nothing then, but in later years admitted resenting the scrutiny. Tensions mounted when, in lieu of affordable sitters, she resorted to bringing her children to the observatory as she worked. Colleagues rolled their eyes when the rambunctious kids ran up and down the spiral staircase among the plate stacks and played with the creaky dumbwaiter. Their father's office was relatively accessible, but to reach their mother's on the top floor of Building D it was necessary to "run the gauntlet of all the open office doors of the scientists," including the director's. Colleagues voted to make their displeasure known. She felt insulted and embarrassed in ways Sergei Gaposchkin likely didn't: in the eyes of her peers the unruly children were the result of their neglectful mother, not their overly indulgent father.[53]

Women of Annie Cannon's generation relinquished traditional family life for a long-term, single-focused commitment to the observatory, but Payne brazenly tried to have it all. Remarkably, the matronly Cannon came to her defense. Professional astronomy need not conflict with a woman's married life, she argued: "Research, which is not confined to fixed hours or necessarily to office walls[,] may easily be carried on by married women. A stellar photograph may be studied at home, during odd hours, and perhaps may not require more time from a wife or mother than is frequently given to bridge playing or various other social activities." She reminded critics, too, that since the establishment of the Pickering fellowships in 1916, ten of thirteen former female recipients, Payne included, continued to be engaged in scientific research or teaching.[54]

In truth, Cecilia Payne-Gaposchkin had been a good investment over the years, well worth her modest salary and then some. Shapley burdened her with excessive teaching and a disproportionate amount of "departmental housework" to free up her male peers for research. She worked at the observatory six days a week, staying often at night for lectures, seminars, and parties, since she was charged with fund-raising and social organizing. She had reason to fear that the added responsibilities made her look scattered, less focused; when it came to promotion she was passed over again and again for men who appeared to be innovating around the clock. When she saw what was going on she told Shapley to put her back on spectroscopy, but he offered spectroscopy positions to outside male candidates and relegated her to photometry so as not to step on their toes. In 1944 when she was passed over for a full professorship, administrators cited her "domestic situation" as the reason. She had heard the same explanation when denied an executive position in the American Astronomical Society in 1938.[55]

James Conant replaced Lowell as Harvard's president and promoted both Cecilia Payne-Gaposchkin and Annie Cannon as full-fledged "astronomers" in 1938, but Payne couldn't hope for further advancement so long as Shapley remained the director. Donald Menzel, who finally replaced Shapley in 1954, was shocked to discover in employment records that his able female colleague had been paid a fraction of his salary over the years. Immediately he set procedures in motion to double her salary, and shortly thereafter promoted her to a full-tenured professorship, followed by a department chair. Surprisingly, she was not particularly joyous upon hearing the news, for, as her daughter recalled, the position of chair was a source of anxiety that her mother "combated with cigarettes, coffee and an occasional pill." Payne was then almost sixty and ready to relinquish all administrative responsibilities to focus on research. The upside, of course, was that the formal university appointment increased her salary by a factor of four.[56]

Annie Cannon shared Cecilia Payne-Gaposchkin's international acclaim, and women in succeeding decades have shared her academic credentials, but no one had had both during her five decades at the observatory. Her status was unique, and thus also were her challenges as a mother, wife, and scientist. Younger women were grateful that she took many of the blows pioneers suffer to clear paths for others. Dorrit Hoffleit, the fifth woman astronomy doctorate holder at Harvard in 1938, had been shepherded into work on the light curves of variable stars (at the improved rate of forty cents an hour), before leaving during World War II to work as a missile trajectories expert at the Aberdeen Proving Ground. Similarly Henrietta Swope doubled as independent scholar and assistant to Shapley before moving on to research positions in the radiation labs of MIT and the navy's hydrographic division. One wonders if Payne would have been promoted sooner had she leveraged similar wartime positions early in her career.

Nevertheless, when the war ended, Hoffleit and Swope found themselves once again in "female" niches within astronomy. Swope took an associate position at Barnard and left in 1952 to assist in an investigation of variable stars of the Andromeda Nebula at Mount Wilson Observatory. Her supervisor, Walter Baade, assured her that she would be a partner of equal rank, but in the end administrators classified her as "a B.A., female computer." Swope was discouraged, but not surprised. "I have always held an innocuous title," a "research assistant" at Harvard, a "staff member" at MIT, and a "mathematician" in the Navy. She was proficient at classifying stars, but male astronomers found it hard to classify a woman with her skills and credentials. She understood what Payne must have felt for decades under Shapley.[57]

Swope remained out west, but was denied access to the telescope at the Palomar Observatory until the mid-1960s. Ultimately a telescope was named in her honor, much as awards were named after Annie Cannon. But her career path reveals the contours of the story told in different ways through the careers of Cannon, Payne-Gaposchkin, and the others: although astronomy was one of the first science fields to open professional opportunities to women, it simultaneously denied them and left them ghettoized in low-status work. Leaving home did not help the female pioneers of the field shed their associations with domesticity. Even as they engaged in a masculine brand of scientific inquiry, professional men interpreted their work as a sort of housekeeping among the stars.

Notes

1. Anonymous astronomer quoted in Patricia Ann Palmieri, *In Adamless Eden: The Community of Women at Wellesley* (New Haven, CT: Yale University Press, 1995), 86.

2. The script sat for decades until the staff put on the production in 1929. Typed script of observatory production of "The Observatory Pinafore," 1929, in Henrietta Swope Papers, Carton 2, folder 89, Schlesinger Library, Radcliffe College, Cambridge, MA.

3. "Good Housewives Will Understand," *Camden Daily Courier*, n.d., in "Clippings, Draper Medal 1931," Biographical Clippings, Annie Jump Cannon Papers, Harvard University Archive, Cambridge, MA.

4. M. E. Cannon to Annie Jump Cannon, September 14, 1880, Personal Correspondence, 1873–1941, A–P, Box 1 of 2, Annie Jump Cannon Papers; Edna Yost, *American Women of Science* (Philadelphia: Frederick A. Stokes, 1943), 28–29, 33–36.

5. Margaret W. Rossiter, "'Women's Work' in Science, 1880–1910," in *History of Women in the Sciences: Readings from Isis*, ed. Sally Gregory Kohlstedt (Chicago: University of Chicago Press, 1999), 292–304.

6. George Johnson, *Miss Leavitt's Stars: The Untold Story of the Woman Who Discovered How to Measure the Universe* (New York: W. W. Norton, 2005), 15–17; Pamela E. Mack, "Straying from Their Orbits: Women in Astronomy in America," in *Women of Science: Righting the Record*, ed. G. Kass Simon and Patricia Farnes (Bloomington, IN: Indiana University Press, 1993), 72–73.

7. Mack, "Straying from Their Orbits," 85, 87.

8. George Greenstein, *Portraits of Discovery: Profiles in Scientific Genius* (New York: Wiley, 2001), 25; Londa Schiebinger, "Maria Winkelmann at the Berlin Academy: A Turning Point for Women in Science," in *History of Women in the Sciences*, 58.

9. Rossiter, "'Women's Work,'" 289–90; Laurence Goodman, "Pickering's Harem," (thesis, Harvard University, 1989), 5–6, Harvard University Archives; Johnson, *Miss Leavitt's Stars*, 19; Mack, "Straying from Their Orbits," 76–79, 86, 89.

10. Mrs. M. Fleming, "A Field for Woman's Work in Astronomy," reprinted from *Astronomy and Astro-Physics*, no. 118 (read at the Congress of Astronomy and Astro-Physics, Chicago, August 1893), in "Clippings on Williamina Paton Fleming, HUG 1396.5, Harvard University Archives, 3; Mack, "Straying from Their Orbits," 89; Rossiter, "Women's Work in Science," 291.

11. Annie Jump Cannon, "Astronomy for Women" for the "Every Evening," February 7, 1929, Speeches and other manuscripts, Box 2 of 2, Annie Jump Cannon Papers; G. Kass Simon, "Biology Is Destiny," in *Women of Science*, 227–30.

12. Annie J. Cannon, "Williamina Paton Fleming," *Science* 33, no. 861 (1911): 987–88, in "Clippings on Williamina Paton Fleming," Williamina P. Fleming Papers, Harvard University Archives; Yost, *American Women of Science*, 39; Cecilia Payne-Gaposchkin, "The Dyer's Hand: An Autobiography," Typed manuscript, Schlesinger Library, Radcliffe College, 54–55; Mack, "Straying from Their Orbits," 100.

13. Johnson, *Miss Leavitt's Stars*, 86; Entries for March 2, 7, in "Journal of Williamina Paton Fleming—Curator of Astronomical Photographs, Harvard College Observatory," Harvard University Archives.

14. Examples of popular works by Clerke are *A Popular History of Astronomy During the Nineteenth Century* (1885), *The System of Stars* (1890), *The Herschels and Modern Astronomy* (1895), *Astronomy* (1898), *Problems in Astrophysics* (1903), and *Modern Cosmologies* (1905). Bernard Lightman, "Constructing Victorian Heavens: Agnes Clerke and the 'New Astronomy,'" in *Natural Eloquence: Women Reinscribe Science*, ed. Barbara T. Gates and Ann B. Shteir (Madison, WI: University of Wisconsin Press, 1997), 61–65; C. E. Barns to Annie Jump Cannon, April 4, 1926; December 22, 1929, Personal Correspondence, 1873–1941, A–P, Box 1 of 2, Annie Jump Cannon Papers.

15. H. H. Turner to Miss Cannon, April 29, 1925, Oxford Degree, 1925, folder, Professional Correspondence and other papers, 1883–1941; Annie Jump Cannon to Miss Katharine A. Jones, Chairman of Program, Chicago Woman's Club, April 30, 1921; May 1933 Bulletin of the Chicago Woman's Club, Chicago Woman's Club folder, Professional Correspondence and other papers, 1883–1941, Box 2 of 4, Annie Jump Cannon Papers.

16. Harriet Bigelow to Annie Jump Cannon, June 5, 1933; Anne S. Young to Annie Jump Cannon, May 19, 1933; Louise S. McDowell to Margaret Ferguson, April 18, 1933; Helen S. French to Margaret C. Ferguson, April 18, 1933; Margaret C. Ferguson to Annie Jump Cannon, April 21, 1933, in Women in Science Chicago Speech, Hall of Science, June 20, 1933 folder, Speeches and other manuscripts, Box 1 of 2, Annie Jump Cannon Papers.

17. Typescript of "Women in Science, 1833–1933," in "Women in Science Chicago Speech, Hall of Science, June 20, 1933" folder.

18. Annie Jump Cannon, "Astronomy for Women."

19. "A Report of the Methods & Results of Meteor Observing"; A. Grace Cook to Annie Jump Cannon, February 9, 1921, A. Grace Cook Folder, 1920–21; "The Annie J. Cannon Prize," Professional Correspondence and other papers, 1883–1941, Box 1 of 4, Annie Jump Cannon Papers.

20. "Mrs. Williamina P. Fleming Dead," *Boston Globe*, May 22, 1911; Cannon, "Williamina Paton Fleming"; Mack, "Straying from Their Orbits," 92–94; Johnson, *Miss Leavitt's Stars*, 20.

21. Annie J. Cannon, "Mrs. Fleming," *Scientific American*, June 3, 1911, in "Clippings on Williamina Paton Fleming."

22. Garret P. Stevens, "A Woman's Achievements in Astronomy," *Chicago Examiner*, June 29, 1911, in "Clippings on Williamina Paton Fleming."

23. Entry for March 10, 1900, "Journal of Williamina Paton Fleming."

24. Entries for March 1, 4, 9, 10, 11, 12, in "Journal of Williamina Paton Fleming."

25. Entries for March 12, 5, 3, in "Journal of Williamina Paton Fleming"; Goodman, "Pickering's Harem," 10.

26. Edward C. Pickering, "Henry Draper Memorial," in the Annual Reports of the President and Treasurer of Harvard College, 1902–1903, 253, electronic version, Harvard University Archives; Louise W. Carnegie to Williamina Paton Fleming, October 6, 1908, Miscellaneous Correspondence, 1905–1909, Williamina Paton Fleming Papers; Entries for March 4, 14–31, in "Journal of Williamina Paton Fleming."

27. Stevens, "A Woman's Achievements in Astronomy"; Edward C. Pickering, In Memoriam, "Williamina Paton Fleming," Williamina Paton Fleming Papers.

28. Mack, "Straying from Their Orbits," 94–97.

29. Goodman, "Pickering's Harem," 11–13.

30. Goodman, "Pickering's Harem," 13–14; Mack, "Straying from Their Orbits," 96.

31. Cannon classified spectra at the rigorous rate of more than three per minute. See Mack, "Straying from Their Orbits," 100.

32. Goodman, "Pickering's Harem," 11–12, 14; Mack, "Straying from Their Orbits," 96–97; Payne-Gaposchkin, "The Dyer's Hand," 48, 54.

33. Johnson, *Miss Leavitt's Stars*, 22–29, 34.

34. Jean L. Turner, "Henrietta Swan Leavitt," in *Out of the Shadows: Contributions of Twentieth-Century Women to Physics*, ed. Nina Byers and Gary Williams (New York: Cambridge University Press, 2006), 56–65; Johnson, *Miss Leavitt's Stars*, 36–38, 41–43.

35. Johnson, *Miss Leavitt's Stars*, 118–19, 90; Mack, "Straying from Their Orbits," 104.

36. Peggy A. Kidwell, "Cecilia Payne-Gaposchkin: Astronomy in the Family," in *Uneasy Careers and Intimate Lives: Women in Science, 1789–1979*, ed. Pnina G. Abir-Am and Dorinda Outram (New Brunswick: Rutgers University Press, 1989), 224; Katherine Haramundanis, "A Personal Recollection," in *Cecilia Payne-Gaposchkin: An Autobiography and Other Recollections* (New York: Cambridge University Press, 1996), 40–41.

37. Payne-Gaposchkin, "The Dyer's Hand," 7–17, 18.

38. Peggy Aldrich Kidwell, "Women Astronomers in Britain, 1780–1930," in *History of Women in the Sciences*, 221–33; Payne-Gaposchkin, "The Dyer's Hand," 26–36.

39. Peggy A. Kidwell, "An Historical Introduction to 'The Dyer's Hand,'" in *Cecilia Payne-Gaposchkin*, 13.

40. Payne-Gaposchkin, "The Dyer's Hand," vi, 47–49, 52, 109, 110.

41. Harlow Shapley to Gerard Swope, May 14, 1927, Carton 1, folder 17, Henrietta Swope Papers; Kidwell, "Cecilia Payne-Gaposchkin," 221–22.

42. Payne-Gaposchkin, "The Dyer's Hand," 111; Kidwell, "Cecilia Payne-Gaposchkin," 223; Vera Rubin, "Cecilia Payne-Gaposchkin," in *Out of the Shadows*, 164.

43. Rubin, "Cecilia Payne-Gaposchkin," 159; Greenstein, *Portraits of Discovery*, 16; Kidwell, "An Historical Introduction," 19; Payne-Gaposchkin, "The Dyer's Hand," 70.

44. Kidwell, "An Historical Introduction," 22–23; Payne-Gaposchkin, "The Dyer's Hand," 111, 61–69.

45. Cecilia Payne to H. N. Russell, December 11, 1930, H. N. Russell Papers, Princeton University Library, Princeton, NJ (letter cited in Kidwell, "An Historical Introduction," 26).

46. Kidwell, "An Historical Introduction," 27.

47. Payne-Gaposchkin, "The Dyer's Hand," 88–92; Virginia Trimble, "An Introduction," in *Cecilia Payne-Gaposchkin*, xix; Kidwell, "Cecilia Payne-Gaposchkin," 226–28.

48. Greenstein, *Portraits of Discovery*, 17–18; Payne-Gaposchkin, "The Dyer's Hand," 94; Kidwell, "Cecilia Payne-Gaposchkin," 229.

49. Irene Curie-Joliot to Missy Meloney, March 29, 1932, June 27, 1937; Dorothy Dunbar Bromley, "Two Who Carry on the Curie Tradition," *New York Times*, January 1, 1933, SM4, Box 2; William B. Meloney-Marie Curie Special Manuscript Collection, Columbia University Libraries, New York, NY.

50. Kidwell, "Cecilia Payne-Gaposchkin," 229–235; Haramundanis, "A Personal Recollection," 64; Greenstein, *Portraits of Discovery*, 22–23; Payne-Gaposchkin, "The Dyer's Hand," 93.

51. Haramundanis, "A Personal Recollection," 39–44, 64.

52. Haramundanis, "A Personal Recollection," 45, 64; Greenstein, *Portraits of Discovery*, 22–23; Kidwell, "Cecilia Payne-Gaposchkin," 234.

53. Haramundanis, "A Personal Recollection," 47–48; Kidwell, "Cecilia Payne-Gaposchkin," 232–33.

54. Report of the Astronomical Fellowship Committee," February, 1935, Fellowship Miscellaneous letters and papers, 1911–35 folder, Professional Correspondence and other papers, 1883–1941, Box 2 of 4, Annie Jump Cannon Papers.

55. Payne-Gaposchkin, "The Dyer's Hand," 111–12; Kidwell, "Cecilia Payne-Gaposchkin," 234.

56. Haramundanis, "A Personal Recollection," 60; Payne-Gaposchkin, "The Dyer's Hand," 112; Rubin, "Cecilia Payne-Gaposchkin," 164.

57. Dr. Walter Baade to Henrietta Swope, October 16, 1951; Henrietta Swope to Dr. I. S. Bowen, Mt. Wilson and Palomar Observatories, October 24, 1951, carton 2, folder 87, Henrietta Swope Papers.

II THE CULT OF MASCULINITY IN THE AGE OF HEROIC SCIENCE, 1941–1962

O N OCTOBER 2, 1928, THE USS *S-21* EMBARKED ON A MISSION TO determine the shape of Earth. Geophysicists had figured that with proper gravity measurements, they could determine the planet's internal density distribution and come closer to understanding the forces creating Earth's structure. Men of the U.S. Coast and Geodetic Survey had proposed a theoretical construct, but they lacked data from the ocean basins. The development of a pendulum gravimeter made them newly optimistic about getting their measurements at sea, and with funding from the Carnegie Institution of Washington, the U.S. Navy recruited adventure seekers to take on the perilous expedition.

When the USS *S-21* returned, its scientists enjoyed state dinners, medals, and media coverage. It was the first time an American submarine had obtained gravitational data, forever changing geologic understanding of the planet. *Scientific American* featured a spread about the research of the expedition and many of the scientists involved. All their findings were kept in an official report that included an appendix on computational procedures and calculations for each of the fifty gravity stations of the expedition, yet the scientist responsible for these data and procedures, a navy astronomer named Eleanor Lamson, was nowhere to be found in the literature memorializing the event.

Like her contemporary Annie Cannon, Lamson had a master's degree in astronomy and had moved up the ranks slowly, beginning as an entry-level computer, becoming head of a computing section, and earning a promotion to associate astronomer. She had paid her dues, and though she had always earned less than men, some recognized her skills enough to call her a scientist. Why, then, has she been remembered as a supporting member of the USS *S-21* expedition when she has been remembered at all?

Lamson's crunching of data from images resembled the computational work of her contemporaries at the Harvard Observatory.

Whether determining the dimensions of the galaxy or of Earth itself, such empirically objective work—statistical measurement from highly complex data that supposedly didn't need interpretation at every turn—was consistently viewed as the output of female technicians rather than of scientists. But this is only part of the reason for her obscurity. When Lillian Gilbreth and the Harvard computers occupied scientific space described as invariably male, their historical obscurity could be blamed on the stereotypical classification of women's appropriate sphere. But as Naomi Oreskes has explained, Lamson's obscurity needs to be seen in a still broader context, as the inevitable effect of a masculine construction of heroism that excluded women, even when particular women embodied heroic traits.

Press coverage of the geologic expedition invoked metaphors of military conquest and athletic competition in a tale of adventure on the high seas. These metaphors presupposed masculine endurance, competitiveness, physicality, bravery, and grit. Sports enthusiasts, explorers, conquerors, and scientific discoverers had been imagined as male heroes, thus hardly evoking in the American mind a portrait that included Eleanor Lamson. In the epic of the USS S-21, a woman could have been imagined as a hero's companion or a loved one pining for her man's return. Lamson was neither, and hence virtually ignored.[1]

In the decades between the world wars the science hero appeared in several manifestations, some more adventurous, cerebral, and eccentric than others. Albert Einstein, a German Jew, became the most recognizable face of science in the United States during the 1920s and 1930s. His celebrity resulted from his theory of relativity (as far as it could be popularly comprehended), but also from the heightened prestige of physics that followed the development of quantum mechanics. He represented "pure science," the science theorized in the mind, rather than applied science, which had once seemed worthy of greater esteem. The recipient of an endless stream of awards and honorary degrees, Einstein, the shy genius, attracted politicians, academicians, and curious bystanders during his U.S. tours. To Americans he embodied the democratic spirit: he seemed to be unimpressed with pretense, and his casual physical appearance seemed to indicate that his mind was totally engaged in abstract thought. He had become an icon of a brand of genius—a form of Platonic transcendence, really—that was decidedly male.[2]

Women's relationship to genius was as caretaker of the bodily needs of those men who had achieved it, since women were too distracted by domesticity to achieve genius themselves. In Einstein's young life, the caretaker had been his wife, Mileva Maric, who gave up her own science career to help his and to care for his children. The American press said little about her, thus making its coverage of Einstein selective in ways it wasn't for Marie Curie during these same years. Sordid details of Curie's love affair, seen fit to print, had adversely influenced her candidacy for an honorary degree at Harvard. While Einstein, in the same period, sired a child out of wedlock and left his wife for another woman, his image as a saint of science could not be tarnished. Curie was occasionally accused of being neglectful of her children as she focused so intently on her research; male scientists, by contrast, could be revered for their single-minded devotion to their work. It was irrelevant that such devotion wasn't in evidence at home.[3]

Einstein had captured the American imagination, and other male physicists, by association, did so as well. Newspapers subscribed to the Science Service to such an extent in the 1930s that news of developments by all scientists reached about one-fifth of the reading public. Yet, in a study of the most oft-cited scientists in mainstream magazines from the 1910s to 1950s, physicists such as Robert Millikan and Michael Pupin appeared more often than all scientists in other fields combined. Surprisingly, Margaret Mead also made the list, although less surprisingly, her science was the "softer" field of cultural anthropology, deemed better suited to women's skilled observation of domestic contexts.[4] Medicinal science, nutrition, hygiene, sanitation—even chemistry— were seemingly "softer," too, in the afterglow of abstract physics, especially because they could be linked to domestic science and its spheres of concern. It was not a stretch to equate the mixing of chemicals in precise measurements to the mixing of flour and sugar for cake baking, but the theoretical thinking required of physicists seemed to have no domestic correlate.[5]

Earlier in the twentieth century Marie Curie, Lise Meitner, Ida Noddack, and Harriet Brooks had had relative success in the "hard" field of physics, but they had not shifted its masculine cast. In the 1920s and 1930s students at women's colleges followed the lead of these women physicists and majored in physical science in proportions equal to men, but their numbers tapered off in the graduate ranks. Four times more women received doctorate degrees in the biological and social sciences in the 1930s than in physics, a field in which men earned 97 percent of the doctorates and won an even greater percentage of research posts afterward. Administrators cited women's lack of experience in prestigious research institutions as one factor, their tendency to marry and leave the profession as another. But Lamson's invisibility also testified to the major problem of perception: few people could imagine women as physicists. This cultural factor cannot be minimized, for it has led to the historical obscurity of women who became physicists, despite the challenges, and to other women's internalization of the belief that they could never be monastic, cerebral, or masculine enough to enter the field.[6]

The rare woman physicist stagnated at women's colleges and small teaching institutions until World War II opened opportunities for study, teaching, and research in major universities. In engineering, one of the few fields proportionately more male than physics, twenty-nine graduate programs opened doors to women during the war, and in all science fields university administrators temporarily rescinded antinepotism policies that, more often than not, barred female job applicants over male ones. Women taught in 12 percent of science classrooms in 1942 and nearly 40 percent by 1946; in chemistry and biology women on college faculties increased by nearly 230 percent. Women also made temporary inroads in weapons, radar, and other industries that were once restricted to men. Historically, wars have had this loosening effect, creating states of emergency that relax gender roles and open opportunities in masculine fields once closed to women. As men left their science jobs to fight overseas, women such as Dorrit Hoffleit, the underpaid computer we met in chapter 3, discovered that her PhD in astronomy allowed her to become a ballistics expert at Aberdeen Proving Ground. Many of the women in the pages that follow—Rosalyn Yalow, Maria Goeppert Mayer, Chien-Shiung Wu, and Rosalind Franklin— jump-started their careers in the midst of wartime contingencies.

Women meteorologists, geologists, and oceanographers joined the navy's Hydrographic Office, the Military Geology Unit of the U.S. Geologic Survey, the Weather Bureau, and the Bureau of Standards, while mathematician Grace Hopper, a navy lieutenant, programmed Mark I, the first large-scale computer, for the Bureau of Ordnance. Women found masculine fields porous during the war, while those working in female niches of expertise became culturally more relevant than ever before. Members of the National Research Council's

Food and Nutrition Committee created guidelines for wartime diets and rationing, and women of the Office of Scientific Research and Development (OSRD) found new uses for penicillin and antimalarial drugs. Government officials revisited Lillian Gilbreth's research on efficiency, both at home and in industry, as homemakers struggled to work a second shift and get supper on the table. While the numbers of women getting science credentials and jobs increased significantly during the war, Margaret Rossiter cautions that the war did not create a qualitative sea change: the percentage of female to male scientists on the National Roster was virtually unchanged (4 percent in 1941 versus 4.1 percent five years later), and most women were to be found in low-level, temporary, and untenured positions, overqualified for their jobs. Many women with science PhDs worked for the OSRD as librarians and technical aides. A male chemistry graduate student at Berkeley recalled being a wartime shift supervisor of women, who, with their greater experience, taught him proper experimental techniques. His experience was not unusual.[7]

Americans knew little about them, but there were also women with the skill set needed to develop the ultimate of wartime technologies: the atom bomb. President Franklin Roosevelt had become fixated on it in 1939, when Einstein warned that nuclear chain reactions were possible and powerful bombs could be produced in Germany if not in the United States first. FDR created federal offices and committees to organize research and development, and soon university labs were part of the Manhattan Project. When all scientific work related to the bomb eventually merged under Robert Oppenheimer at Los Alamos, women worked in his midst. Indeed, they were waitresses, clerks, teachers, and nurses at project sites, as well as computers, technicians, and high-level scientists. Yet when the bombs went off and the war had ended, male scientists—physicists in particular—were the greatest beneficiaries of the postwar spoils. A study of public perception indicated that all scientists garnered prestige after the war but that the nuclear physicist (presumed male) ranked highest, moving from eighteenth out of ninety in 1947 to third by 1963, just behind high-ranking politicians and diplomats. A writer for *Life* magazine told readers that these men donned the "tunic of Superman." Robert Oppenheimer, Enrico Fermi, and Edward Teller graced magazine covers, popular and scientific, and their connections to agencies such as the Atomic Energy Commission, the National Institutes of Health, and the National Science Foundation propelled them further into the limelight as authorities on national policy. Americans viewed their scientific genius as the ultimate defense against Red Spread in the Cold War era.[8]

Social scientists, fascinated to know how these great men thought and worked, studied them obsessively. Psychologist Anne Roe published in 1953 a study of sixty-four elite scientists, perpetuating idealized images of the best as male. She concluded that a range of circumstances—being oldest children, losing a parent, coping with physical disability or illness, being orphans—caused these elite men to crave independence and master self-sufficiency early in their lives. While their sisters played with dolls and tea sets, these men hypothesized about the physical world and became worlds within themselves. While their male peers dated, they avoided girls and people altogether. David McClelland later elaborated on Roe's findings, describing elite scientists as loners who typically staved off dating until late in college and married the first girls they courted, their libidos heterosexual but not vigorous, since science was foremost on their minds.[9]

The archenemy of the successful scientist, according to one of Roe's subjects, was a wife who "expect[ed] him home every evening at five or five-thirty." It is impossible, he told her, "to turn out a properly trained and qualified PhD in physics with only four years of graduate work if this is on a 40-hour week." More often than not the male scientist eventually became a father, but he assumed that his family and his life as a scientist necessarily competed, that the emotional energy a wife demanded could rob him of the intellectual energy he needed for research. No woman would work in the same manner, he thought, since no woman could have his physical stamina, single-mindedness, and emotional detachment. Roe's interviewees underscored the belief that scientific virtuosity involved *theoretical* work in the fields of physics first, chemistry second. As one of her subjects put it, "Scientists reach[ed] distinction earlier, more or less in the ratio of the relative importance of intellectual effort to the experimental work." A theoretical physicist can come up with an abstract idea quickly, he explained. "In biology," however, "you essentially have to make experiments, you can't speculate about what an animal is doing and this is hard and takes a long time."[10] In few words, he had explained why his brand of physical science bred greatness sooner, and perhaps why women had succeeded more often in experimental and biological fields: these were not areas of virtuosity, but of insight after sustained work.

On average, Nobel laureates in literature were past their child-rearing years when they wrote their prize-winning works, for in the humanities, wisdom and style came only with years of lived experience. Not true of scientists, however, and theoretical physicists in particular, whose flashes of insight were expected to come in their twenties and thirties. Most Nobel laureates in science published before turning twenty-five and were promoted to full professor by thirty-five. Einstein was twenty-six when he came up with the theory of relativity, and some men who made paradigm-shifting discoveries during the postwar years were even younger.[11] With scientific discovery occurring at a breakneck pace, and with the rapid appearance of published research, the ethic of "keeping up" turned all-consuming. Young scientists reported spending hours daily reading journals, though truly exceptional minds like Enrico Fermi were known to read virtually none of the professional literature, lending credence to the notion that theoretical brilliance was innate—you had it or you didn't, and you knew early on. When Fermi married his wife he vowed he'd retire at forty. "No physicist ever accomplishes anything after forty anyhow," he told her.[12]

That women's biological clocks competed with the tenure clocks and career schedules of male colleagues added fuel to the supposition that women were too distracted for prolonged, high-caliber science. Sidetracked and delayed by child rearing, some women could not advance in the scientific ranks with the same idealized speed as male peers. Nobel Prize-winning biochemist Gerty Cori was fifty-one when she was finally promoted to full professor; physics laureate Maria Goeppert Mayer wasn't hired with pay and tenure until she was fifty-three. It's hard to know whether their promotional ladders were lengthened because they were mothers or women, although given the institutional chauvinism of postwar science, it's likely that the responsibilities of motherhood and the stigma of being female were mutually reinforcing. Regardless of how much children actually interfered, male colleagues perpetuated the idea that home life and science were two distinctly different and incompatible realms.[13]

Roe's study betrays a deeper and broader paradox: while women scientists had seized research opportunities and saw new ones ahead in burgeoning industrial and academic sectors, the stigma attached to

them as domestic and hence "nonscientific" beings cut their opportunities. A study Margaret Mead and Rhoda Metreaux conducted of high
school students in 1957 confirmed the discrepancy. When asked to describe their image of the ideal scientist, students decided he was intellectual and cold. If a forensic illustrator were to draw this figure, he
would look bearded, bespectacled, and decidedly virile. Male and female respondents imagined similar portraits and assumed that women
could be the wives of scientists, not scientists themselves. Such caricatures conformed to the image vivified by Roe, but also by parents,
teachers, Cold War propagandists, human resource departments, television, and sci-fi films. It was no coincidence that the majority of science fiction readers in the 1950s were male college graduates under the
age of thirty-five and already engaged in technical occupations: they
most closely resembled and identified with the comic book heroes who
saved the world from nuclear and biological annihilation.[14]

The sudden cultural amnesia that blotted out women's contributions
to wartime science is striking. In 1943 women were enticed to enter labs
through cinematic images of Marie Curie; five years later, everything
from female enrollment quotas to reinstated antinepotism policies and
social stereotypes deterred women from living out the Curie dream.
The pervasive message in the 1950s was that science was not for them,
and they internalized it thoroughly. Only one-fourth of women college
graduates wanted professional careers, with science ranking low on
their preferred list of fields. Recruiters at Boeing Airplane maintained
that they would hire women engineers if they could find them, and yet
Radcliffe's class of 1957 included only three graduates qualified to take
on scientific work of any kind. Quietly, personnel directors thought it
just as well. One admitted that he simply couldn't imagine a woman
walking the grounds of his cement plant, and others at steel companies summarily denied the applications of women engineers. "It would
be pretty hard to send a woman out with a geophysical field survey
party that might be out in the field for months," a Gulf Oil recruiter
explained.[15]

Of course there were women who had been scientists during the
war and wanted to be afterward, but in this milieu opportunities dried
up and went to male beneficiaries of educations paid for by the GI Bill.
Mathematician Naomi Livesay tried for a time to resume a career in
England after leaving Los Alamos, but found that her low salary didn't

justify the child-care expense. Clinton Engineering Works chemist Ellen Weaver got a doctorate in biology but struggled with a graduate advisor who disapproved of her having babies and thus refused to attend her oral exams. "There was a strong feeling that women should be at home," she wrote of the 1950s. "The general attitude of the family and other people was, 'Well the war's over, so why are you bothering with this?'" In nearly all sectors of wartime science men replaced women and enjoyed smaller teaching loads, better pay, and more professional esteem in these positions. The number of men who earned science PhDs between 1946 and 1960 skyrocketed from eight hundred to four thousand a year. Never in these years did the number of women PhDs reach five hundred annually.[16]

In some ways women physicists suffered most severely, since the forces driving women out or to the margins were greater in the most prestigious fields. Female physics PhDs dropped to one in forty in the 1950s, and proportionally fewer female than male physicists found industrial jobs after graduate school. Robert Millikan had stated back in 1936 that women showed no greatness in the "father" of all science fields. Other than Lise Meitner and Marie Curie, he could think of no woman—certainly not an American one—with the talent to succeed. Hence, teaching was where women belonged. During the war Chien-Shiung Wu and Maria Goeppert Mayer had proved him wrong, but their individual success did nothing to change pervasive attitudes about women, physics, and pedagogy. Elda Anderson, a seasoned physicist who had worked at Los Alamos for four years, could find only teaching positions after the war, one at Downer's College and one as a health educator at Oak Ridge.[17]

If women scientists wanted to remain in the workforce, often their only options were in sectors that required no lab or fieldwork at all. Teaching was one option, but recruiters for the chemical industry also encouraged women to consider work as "chemical secretaries," science editors, and librarians. They insisted that women were organized, neat, good-natured, even-tempered, and especially talented at mundane tasks that would bore men with similar credentials. Applied to lab work, such traits translated into the work of technicians or those responsible for quality control. Thus, women became the majority working in the merchandise control lab at Sears, Roebuck: "Who better to test cosmetics and textiles?" inquired one recruit. Pockets of women employed in research and development could be found in food science industries, in nutritional chemistry, among medical technicians, and in technician jobs generally. One estimate holds that four out of every five technicians were women in 1940, and percentages remained steady through the postwar years.[18]

In sum, women scientists through the 1940s and 1950s hovered in low-ranking, temporary, and pedagogical positions in fields that had suffered diminished prestige after the war. Science recruiters in the 1950s preferred to hire men rather than women, and in fact proportionally more men trained for jobs in science fields. Low birthrates in the depressed 1930s and the rapid growth of technological sectors after the war did force recruiters to consider hiring some women nonetheless. And though they urged girls to take math and lab courses to compete with technologically savvy Soviet women, they also reminded them that such women were not attractive to men. Not surprisingly, the paradox persisted: actual numbers and needs could not eliminate or even soften cultural perceptions about the gender of skilled laborers or American scientists. The result, according to historian Margaret Rossiter, was cultural blindness: "Except for the ubiquitous Margaret Mead and the outspoken Rachel Carson, women scientists were practically invisible to the public, to other scientists, and even to each other in the 1950s and 60s."[19]

In 1928 myth had made it impossible for journalists to envision Eleanor Lamson as part of the USS *S-21* expedition. Thirty years later this myth and others eliminated science careers for women before they had begun. In chapter 4 we see how mythologies obscured the lived realities of the Manhattan Project, and in chapter 5 we tease myth from reality in the careers of Maria Goeppert Mayer, Rosalind Franklin, and other women who occupied labs in the postwar years. Women may have been impossible science heroes, but that never made them impossible scientists.

Notes

1. For analysis of Lamson and scientific heroism, see Naomi Oreskes, "Objectivity or Heroism? On the Invisibility of Women in Science," *Osiris* 2nd ser., 11 (1996): 87–113.

2. Londa Schiebinger, *Has Feminism Changed Science?* (Cambridge, MA: Harvard University Press, 1999), 72–74; Daniel Kevles, *The Physicists: The History of a Scientific Community in Modern America* (Cambridge, MA: Harvard University Press, 1987), 175–77, 212, 269; Margaret Wertheim, *Pythagoras' Trousers: God, Physics, and the Gender Wars* (New York: W. W. Norton, 1997), 187.

3. Jurgen Renn and Robert Schulmann, eds., *Albert Einstein/Mileva Maric: The Love Letters* (Princeton, NJ: Princeton University Press, 1992), xi–xxviii.

4. Kevles, *The Physicists*, 269; Marcel C. LaFollette, *Making Science Our Own: Public Images of Science, 1910–1955* (Chicago: University of Chicago Press, 1990), 51.

5. Alice Rossi, "Barriers to the Career Choice of Engineering, Medicine, or Science Among American Women," in *Women and the Scientific Professions: The MIT Symposium on American Women in Science and Engineering*, ed. Jacquelyn Mattfeld and Carol G. Van Aken (Cambridge, MA: MIT Press, 1965), 64.

6. Kevles, *The Physicists*, 275–77; Wertheim, *Pythagoras' Trousers*, 12.

7. Schiebinger, *Has Feminism Changed Science?* 30–31; Margaret Rossiter, *Women Scientists in America: Before Affirmative Action, 1940–1972* (Baltimore: Johns Hopkins University Press, 1995), 1–26; Ruth Howes and Caroline Herzenberg, *Their Day in the Sun: Women of the Manhattan Project* (Philadelphia: Temple University Press, 1999), 18.

8. Harriet Zuckerman, *Scientific Elite: Nobel Laureates in the United States* (New York: The Free Press, 1977), 2; Kevles, *The Physicists*, ix, 334; Ralph E. Lapp, *The New Priesthood: The Scientific Elite and the Uses of Power* (New York: Harper and Row, 1965), 4.

9. Anne Roe, *The Making of a Scientist* (New York: Dodd, Mead, 1953); Evelyn Fox Keller, *Reflections on Gender and Science* (New Haven, CT: Yale University Press, 1995), 91.

10. Roe, *The Making of a Scientist*, 41, 48, 56, 60.

11. James Watson was twenty-five when he discovered the double helix in 1953; Joshua Lederberg was thirty-three when he won his Nobel Prize for a decade's research in bacterial genetics in 1958. See Rossi, "Barriers," 103; Zuckerman, *Scientific Elite*, 145–49, 162, 165–68.

12. Stephen Cotgrove and Steven Box, *Science, Industry, and Society: Studies in the Sociology of Science* (New York: Barnes and Noble, 1970), 1; Walter Hirsch, *Scientists in American Society* (New York: Random House, 1968), 125–26; Laura Fermi, *Atoms in the Family: My Life with Enrico Fermi* (Chicago: University of Chicago Press), 146.

13. Zuckerman, *Scientific Elite*, 192.

14. Margaret Mead and Rhoda Metreaux, "Image of the Scientist Among High School Students," *Science*, August 20, 1957, 384–90; Walter Hirsch, "The Image of the Scientist in Science Fiction: A Content Analysis," *American Journal of Sociology* 63, no. 5 (1958): 507.

15. "Girl Graduates," *Wall Street Journal*, July 1, 1957, 1; "Girl Graduates: Job Opportunities Widen, But Barriers Remain," *Wall Street Journal*, February 1, 1962, 1; Rossiter, *Women Scientists in America, 1940–1972*, 27–49.

16. Weaver quoted in Ruth Howes and Caroline Herzenberg, "Women of the Manhattan Project," *Technology Review*, November/December 1993, 39, 40.

17. Kevles, *The Physicists*, 370–71; Howes and Herzenberg, *Their Day in the Sun*, 59–60, 191–92; Sharon Traweek, *Beamtimes and Lifetimes: The World of High Energy Physics* (Cambridge, MA: Harvard University Press, 1988).

18. Janet D. Scott, "My Work with Chemical Abstracts," 90–91; Cornelia T. Snell, "Writing About Chemistry," 395–96; Margaret D. Foster, "The Chemist in the Water Resources Laboratory," 52; W. S. Landis, "Women Chemists in Industry," 362–72; Elizabeth S. Weirick, "Experiences in the Field of Merchandise Control," 389; Florence E. Wall, "Cosmetics—a Fertile Field for Chemical Research"; Evelyn Hearsey, "Training and Qualifications of the Woman Chemist in the Control Laboratory," 392, in *The Chemist at Work*, ed. Roy I. Grady and John W. Chittum (Easton, PA: Journal of Chemical Education, 1940).

19. Rossiter, *Women Scientists in America*, 64–65, 304.

4 Those Science Made Invisible: Finding the Women of the Manhattan Project

A physicist's wife in peacetime and a physicist's wife in wartime are, I have discovered, two very different things. In the years before our country was at war, this wife's interests were identical with those of any other academic lady. . . . Even before the Pearl Harbor attack, however, the physicist's routine had changed. Defense projects were started in college laboratories; armed guards began to pace the thresholds of physics buildings. One's husband grew more secretive about his work, and one knew that his job must be important, for he was immune from the draft. The physicist's wife realized that her husband, in wartime, was more than just a college professor—his was a key profession in the defense of his country.

—Ruth Marshak, wife of Robert Marshak, deputy head of a theoretical physics group at Los Alamos, 1943–1946[1]

The women on The Hill could not be termed temporary war widows like my sister-in-law in Phoenix, whose husband[,] Preston, was in the South Pacific. We could better be termed camp followers, attendants of the men who made the final assault on Japan. We occupied the sidelines of history and our role was not easy. It was up to us to see that our men were fed and loved and kept serene, so they could give their full attention to the Bomb, the still-winged ant queen who reigned in the Tech Area. We coped with our problems alone.

—Eleanor Jette, wife of Eric Jette, group leader in the Metallurgy Division at Los Alamos, 1943–1945[2]

For the past twenty months you have worked as an assistant in our research laboratory making microscopic measurements which called for a great deal of judgment on your part. This work was exceedingly tedious and involved a good deal of nervous strain. Nevertheless you have performed your duties in a cheerful and diligent manner, and it must be clear to you that you have made a real contribution to the success of the project.

—Robert Oppenheimer to Lyda Speck, Los Alamos technician, 1943–45[3]

IN 1941 RADIOCHEMIST ELIZABETH RONA GREW ANXIOUS AS THE RUSsians and Nazis encroached on her native Hungary; like Lise Meitner and other scientists she had known intimately in Germany, she decided it was time to flee. She received a visitor's visa to the United States, and her loved ones saw her off. "Tell Roosevelt, or ask Einstein to ask Roosevelt, to get America to enter the war against the Germans," a friend told her. But it took time to be heard by an elite scientist in the United States, let alone to secure a job working with one. Her first stop was Columbia University, where Enrico Fermi, Leo Szilard, and other men with whom she had once worked were now settled in labs of their own. She brought greetings from associates abroad, yet most of these men behaved as though they didn't have people or memories in common. Harold Urey later explained that the FBI was watching, since authorities feared she was a spy. Szilard finally took her to dinner to reminisce, but made sure that they met at a dimly lit Chinese restaurant, where it was too dark to be detected.

Rona was jobless for three months before attending the annual meeting of the American Physical Society, where she met theoretical physicist Karl Herzfeld, who found her a teaching job at a women's college in Washington, D.C. She was just getting to know her students when she got a call to do research—something to do with radioactivity, but no specifics could be conveyed over the phone. Her lack of working papers meant that the job would fall through, but soon after she received a cryptic telegram from the Institute of Optics at the University of Rochester: they wanted her methods for extracting polonium. She guessed that the query was related to war work. Certainly, no one in the world was better than she at extracting the substance in large quantities, and apparently someone wanted to stockpile it. The next day a thin man came to the door to give her details. Find a student to help with the manual labor in the lab, he told her, preferably someone oblivious to the science involved in the extraction. Rona chose a woman majoring in French to assist her and proceeded with the highly confidential science. Suddenly her noncitizen status was no longer a barrier. She got security clearance, and the Office of Scientific Research and Development (OSRD) offered to buy the secret to her methods. She told what she knew for nothing. Plutonium plants were built, but she had no idea where they were or that an international group of scientists had settled in the New Mexico desert with designs for the nuclear material they would produce. Many of the men involved were the ones who first had turned their backs on her when she had arrived in New York in 1941.[4]

A homemaker named Eleanor Jette lived with her husband, Eric, a professor of metallurgy at Columbia University, and their ten-year-old son, Bill, whom she tucked into bed one night in 1943. She came downstairs to find Eric smoking his churchwarden pipe: "Gus came to see me today. . . . He's a consultant for a big, super-secret project run by the Army Engineers. . . . Gus wanted to know if I'd be interested in another war job."

The two sat up contemplating the move, which, as far as Eric could guess, was to some place in north-central New Mexico. Eleanor was dubious. Should she quit her job? Eric urged her to wait until just before they'd be ready to leave. "Nobody's supposed to know we're moving except the people who actually move us," he explained. "We can't tell our friends I'm leaving the university. We can't even tell Bill we're going to move. Children his age are considered to be very poor security risks. Our families will have to know that we're moving, but we can't tell them where we're going." Now Eleanor was really dubious.

Looking through a brochure about the "Los Alamos Project" was hardly acceptable preparation for a life in an unnamed place and for an unknown duration. Eric told Eleanor to leave her technical books behind, but then he packed her little volume on the theory of probabilities for himself. What were the "long odds" he referred to as he pulled the book from the shelf, Eleanor wondered, and why did the men in his research group have to have such elaborate health checks before arriving in New Mexico? His behavior perplexed her, but so did the men she met upon reaching the desert. They complained about a glut of well-trained inorganic chemists—what for? One man with whom she dined was at "the top of the high explosives heap"; another, a Nobel laureate for his work on atomic structures. Although her formal science training was "coated with rust," she had suspicions that her husband was out to build a bomb to end all wars. In July 1945 her hunch was confirmed.[5]

On that night Elsie McMillan heard a knock at the door of her home in the New Mexico desert. It was her neighbor, Lois Bradbury, who couldn't sleep either. The two women stayed up in the kitchen talking, every so often checking on the children and looking out the window for signs of stirring in the distance. Little David McMillan woke up wanting a bottle and was sucking contentedly when his mother spotted a flash through the window that lit up the desert sky. Finally, the experiment that had given her so much anxiety was over. Only a few division heads and members of the Trinity team knew when the bomb was scheduled to explode, yet she had suspected it was to be soon when her brother-in-law, Nobel physicist Ernest Lawrence, had arrived at Los Alamos. Over the past two years she had grown accustomed to visits from world-renowned scientists, but her husband had left the house at 3:15 the prior morning in silence. She knew that things would never be the same.[6]

The oppressive air of secrecy made Los Alamos different from any other American community. Ruth Marshak felt like the heroine in a melo-drama, whisked away by a husband who had packed up her things, re-fusing to say anything about their destination. "I felt akin to the pio-neer women accompanying their husbands across the uncharted plains westward . . . into the Unknown." Indeed in some respects she, too, had become unknown, her former identity eradicated as numbers replaced her name on her driver's license and tax forms. She couldn't vote, di-vorce, own a phone or camera, or speak freely. Like all wives of scientists at Los Alamos, she was fingerprinted and subject to background checks and readings of the Espionage Act. When she wished to reach some-one in the outside world, she wrote letters but left them open for army censors, who would reject them or send them to trusted secretaries to postmark.[7]

Santa Fe residents wondered whether Los Alamos was a base of operations for building a Russian submarine or a home for pregnant members of the Women's Army Corps (WACs). Clarity was never forth-coming. Prominent scientists—under aliases—went about with body-guards, as did their families. On their rare shopping excursions to Santa Fe, women could not refer to husbands and colleagues as physicists and chemists. Rather, they were "fizzlers" and "stinkers" in their strange and censored world. Marshak felt as though she were living on an "island in the sky," since fences separated her from the world both physically and psychically. Laura Fermi, whose husband was Enrico Fermi, had had a preview of this kind of oppressive life when her husband had worked at the University of Chicago in 1942. Armed guards patrolled outside the physics building as wives viewed a film called *Next of Kin*, the story of a British officer's loose lips, which had led to the bombing of London. Laura Fermi understood that women, by remaining ignorant about the lab, were to avoid spilling the beans. Months later when she arrived at

Los Alamos with her children, a soldier approached her and asked if she were *the* "Mrs. Farmer," but she had no idea who that was. People wondered later how she could know so little about what was going on, but she insisted that other wives also knew nothing. When Enrico brought up work, she sent him to his office to be alone with his thoughts. Physics wasn't foremost on her mind.[8]

These recollections of experiences during the Manhattan Project don't exist in the historical memory of most Americans. Conjured instead is a secret laboratory in the desert occupied by great men of science, hunkered down without wives, children, or female colleagues to distract them. Such images add further intrigue to an already overly dramatic story of scientific discovery, but they belie the realities of the Manhattan Project. Most historians have treated women as extras and bit players in a drama that features a primary cast of brilliant men. But in fact women's presence casts a revealing light on the project, since they, more than the men, lived at the center of its tensions.

The science of the Manhattan Project took place in the Tech Area at Los Alamos but also in labs at MIT, Minnesota, Rochester, Wisconsin, and other universities throughout the country. Physicist Arthur Compton headed up the Metallurgical Laboratory at the University of Chicago; Harold Urey, the research on gaseous diffusion at Columbia; and Ernest Lawrence, the electromagnetic separation at Berkeley. The DuPont Company oversaw plutonium production in Hanford, Washington, and uranium enrichment plants went up in Oak Ridge, Tennessee. Women worked at all these production and research sites, even if it wasn't what General Leslie Groves, the military overseer of the project, had first envisioned. He wanted the scientific work completed quickly, and he thought that men would leave their wives behind so that they could engage in the no-nonsense work of building a bomb while they lived an isolated life in classified locales. Scientists were supposed to be objectively disinterested, asexual, and emotionally detached, their only urges purely intellectual and patriotic. But Robert Oppenheimer, the physicist Groves hired to head up the scientific divisions of the project, thought it best to accommodate whole families at research sites to keep up morale.

Although Groves organized scientists hierarchically as though they were military men, their civilian spouses, typically wives, humanized and democratized project sites. Single women, too, made an impact: their presence revealed the project as a sexually contested and charged terrain. The military's double standard in its dealing with sexual indiscretion was telling: for single men, promiscuity (if one could call it that) was strictly a health issue; for women, a moral one. The military deemed a woman allowing men in her dorm at Los Alamos an offense as severe as violent crime. The severity of charges against her underscored that she was needed but not wanted on the project. Her femaleness normalized conditions, yet was subversive of the systems of control to which military men were accustomed. Her very presence turned the project into all that it was not supposed to be.[9]

But women were there. Spouses indirectly contributed to the project by freeing their husbands to develop the bomb, and at Los Alamos Pueblo domestic workers also freed wives to free their husbands. But to focus solely on women who supported male scientists would be dismissive not only of WAC volunteers but also of civilian women who worked directly in research and production. Nine percent of the workforce at Hanford was female, and the percentage was higher at Los Alamos. At Oak Ridge, women from the mining towns of the Tennessee mountains became important cogs in the uranium enrichment machine. Around the clock, seven days a week, they sat at their workstations, making sure the gauges on the calutrons stayed within specified limits. Recruiters for these quality control posts looked for a specific profile of worker: women with high school degrees but without a knowledge of science that would be sufficient for them to understand the point of their work. Yet visitors observed that women adjusted the intensity of the magnetic fields on electromagnetic isotope separators more efficiently than did the physicists who had designed them. Because the magnets on the calutrons pulled out their hairpins if they weren't careful, they became masters of the machines.[10]

This image of women operators summons up yet again the Harvard computers, who performed the work of photometry better than did the men who had developed it. Those women were hired to do the task immediately in front them, just as were the thousands of women employed at the Oak Ridge Y-12 plant. Their monotonous tasks forbade their seeing the project's larger purpose, and yet their work tells us a great deal about the project's intentionally gendered compartmentalization of knowledge and tasks. Divisions between thinkers and doers, scientists and laborers, science and domesticity often fell along gendered lines. Even when these divisions were more imagined than real, their impact on the historical memory of the project has been real indeed. As presumed homemakers, women have been compartmentalized out of the same volumes of history that have lionized male thinkers in the lab.

But there were women developing the science of the bomb in labs long before men thought to or cared. These women were pioneering figures in nuclear physics, most of them European, since the field first developed at the Cavendish in Cambridge, the Curie Institute in Paris, and the Wilhelm Institute in Berlin. Harriet Brooks, Marguerite Perey, and Catherine Camie were scientific pioneers at these centers before World War II, while Marie Curie, most notably, introduced radium and radioactivity to the world. Her daughter Irene Curie brought scientists closer to developing the bomb when she discovered artificial radioactivity in 1933. At the same time, the German Ida Noddack, who had discovered rhenium and technetium, found that no one would pay attention to her explanation of the idea of the fissioning nucleus. Otto Hahn, who eventually won the Nobel Prize for the discovery of nuclear fission, had called Noddack's initial suggestion of it "really absurd." His dismissal of her claims was arrogant, but so was his denial of his longtime collaborator Lise Meitner's theorizing of the fission he observed. Personal ego and political agendas are clearly responsible for the omission of women from scientific history, but another factor, usually unaccounted for, is the ethical posture of women themselves. When asked to join the team developing the atom bomb at Los Alamos, Meitner flatly refused, believing it her moral obligation to prevent the mechanisms of war from being realized.[11]

Women have been left out of the history of the Manhattan Project, and yet their views both as insiders and as outsiders offer a complex portrait of the project's science and culture. Some simply accompanied their husbands, but other wives also had their own scientific credentials: among them were Rose Mooney-Slater, Sue Norton, Lilli Hornig, and Jane Hamilton Hall. Other women were promising young physicists such as Kay Way and Joan Hinton, and chemists such as Helen Landriani Bench and Virginia Spivey Coleman. There were also technicians, including Margaret Sanderson and secretaries such as Priscilla Greene Duffield. There were also Hanford mess hall waitress Margaret Hoffarth; Oak Ridge physician Mrs. T. H. Davies; and Los Alamos "oracle" Dorothy McKibbin, who greeted scientists' families when they reached Santa Fe. The list includes also the beloved Edith Warner, who provided the rare civilized dining experience in her tearoom near the Los Alamos labs; the young, single Hope Sloan Amacker, a WAC who doubled as a beauty pageant contestant to keep up morale at Hanford; and even "Ma" Hirschfelder, who came to Los Alamos to serve as house-keeper to her confirmed-bachelor son.[12]

Leona Marshall Libby, who worked as a physicist at the Met Lab in Chicago, in Argonne, and then at Hanford, spoke in later years about the technical facets of the project, although her being female separated her from other scientists and made her experiences fundamentally dif-ferent from theirs. She was a rare woman among men when she received her PhD under Nobel laureates of the University of Chicago's physics department in 1943. Then and later, she was assistant, colleague, swim-ming companion, and wife to some of the most renowned physicists in the world. Libby became part of Fermis' social circle and, at the age of twenty-two, felt somewhat out of place playing "Murder" with elite science couples at their parties: "I shrank into the corner and listened with astonishment to these brilliant, accomplished, famous, sophisti-cated people shrieking and poking and kissing each other in the dark like little kids." Hers was not a historian's perspective, but one more personal and complex.[13]

Coeds might have been found beneath football stands kissing boys, but Libby spent months underneath the Chicago football field perfect-ing a nuclear chain reaction. Per Fermi's request she read the latest the-oretical papers on plutonium-production reactors and prepared coun-ters for the graphite pile she stacked in a remote corner of the basement of Eckhart Hall. Decades later Libby remembered the first controlled reaction under Stagg Field in December of 1942:

> We were nearly ready to leave, taking off our dirty gray laboratory coats and putting on our outdoor coats when in came Eugene Wigner, alone, with one small bottle of Chianti. We were only about 20 people, the visitors and the high brass all having long since left. Eugene had, mi-raculously, not only the tiny bottle of Chianti but also some paper cups. These he filled with a few drips each and passed them around in the midst of that dingy, gray-black surrounding without any word whatso-ever. No toast, nothing, and everyone had a few memorable sips.[14]

Many of the men with whom she drank have since written their own accounts of the event and added to its myth. Few recalled her presence, or at least they failed to mention it. Herbert Anderson gave a nod to his lab partner by including her in a list of witnesses to the event under the ambiguous name "L. Woods." Football players featured in his account of bombmaking as more useful participants, for they packed the tin cans of uranium oxide powder he needed for experiments in 1941. Had Libby's signature not been on the Chianti label, the historical record would have been mute about her work in the significant chain reaction. But she was there, and she became Laura Fermi's trusted informant, since no one else would explain at the party Laura gave that night why her husband was being so gratuitously congratulated.[15]

Indeed Libby had cultivated friendships with both Fermis—albeit on different terms. With Enrico she stayed up until the wee hours talking shop, while Laura bent her ear on novels, the children, even on better and worse brands of olive oil. As a scientist and intimate, she was in a position to see through the myths about her colleagues' lives with perspective and as they were spun. She claimed that Edward Teller, the notorious originator of the hydrogen bomb, was careful to keep himself unpressed, even tangling his thick eyebrows to perfect his introspective look. She held Fermi in high regard as a man rather than as a mythical figure. She noticed his moments of brilliance, but also his quirks; he had revolutionized the study of beta decay, but he also pulled fringe hairs from the back of his neck to floss his teeth. In lionizing biographies such eccentricities have been either ignored or summoned to prove that the men of the project were highly intellectual beings, necessarily without social graces. Libby, however, thought Fermi's eccentricities part of his humanity; from her perspective he was both more and less than an intellectual juggernaut.[16]

Libby's social location as a woman and scientist was also awkward and alienating for her. Like Cecilia Payne-Gaposchkin and others in rarified physics circles, she often distanced herself from the domestic world of women to appear as a scientific professional among peers. Physically, she tried to blend in at work; acceptable attire for men who worked in the Metallurgical Lab was a suit, and hence she wore the closest equivalent, a blazer that sat squarely on her shoulders. She blended in so successfully that her supervisor, Walter Zinn, thought nothing of shouting obscenities around her that he would never utter in mixed company. And yet there were times when male scientists couldn't help but see her as different. When, for example, she and a male colleague were both exposed to large doses of radiation from gamma rays, she was the one to get an earful from the lab doctors, who warned her to protect the limited number of egg cells God had given her. Men looked askance at what seemed to them her lack of appropriate maternal concern, but no one said anything to her male colleague.[17]

In 1943 she married physicist John Marshall, but in her published recollections of these years he became a side note, like all things outside her scientific work, merely a lab partner who left his dirty laundry in her sleeping quarters. She divulged few details about a pregnancy that quickly followed, for it underscored how different she was from men in the lab. She continued to measure neutron cross-sections in the reactor building in preparation for Los Alamos, her colleagues unaware of her condition. "My work clothes—overalls and a blue denim jacket—concealed the bulge, and the pockets, containing side cutters, tape measure, slide rule, micrometer, pen and pencil, needle-nose pliers, small black notebook, and other such essentials, produced other bulges." She returned to work a week after her son was born, since the team needed more measurements before building the plutonium reactors at Hanford. In Washington State, where the reactors were built, female secretaries and food service workers occupied project buildings some thirty miles away; Libby, however, stayed in the thick of things, the DuPont Company building a private bathroom for her in the production plant. She would have relieved herself outdoors if it meant blending in better with her peers.[18]

Libby's mother lightened her domestic burdens by taking care of her grandson at Hanford, but other women at project sites hadn't the same luxury of relinquishing child-care duties to others. Gladys Morgan Happer spent a year at Oak Ridge doubling as a researcher of radioactivity and a clinician during times of radiation exposure; she was also mother and father to a toddler son, one of three children she brought with her to Tennessee. Physicist Elizabeth Graves understood the anxiety of working proverbial double shifts, as well as the fear of exposing unborn children to radioactivity. An expert in fast neutron scattering, she bore three of her five children while engaged in dangerous research. Her decision to be a scientist may have compromised her duties as a protective mother. On the other hand, marriage and motherhood may have injured her psyche as a scientist. Pregnant when Trinity took place, supervisors made her measure radiation further from the epicenter of the blast than the position given to the men in her group.

Colleagues recalled Graves in the lab as a perpetually pregnant presence, who scrambled to finish experiments, on several occasions simultaneously clocking labor contractions. The harried scene is emblematic of what she may have felt as a woman trying to do it all. Another woman of Los Alamos remembered feeling torn as she got caught up in the "hyperthyroid quality" of the Tech Area: "For the potential working wife, there was one chief worry. . . . Could she manage her home here on the mesa and work too? Would her home life suffer? Would her husband be neglected? Would her children become delinquents?" For women scientists the Manhattan Project was often experienced as the fragmentation of self—a perspective rarely seen in history books.[19]

Domesticity and Social Place at Los Alamos

High-minded science and everyday domesticity butted against each other at project sites, particularly at Los Alamos, where the theoretical physics of the Tech Area took place meters from diaper changes in the residences. Mothers complained about army men turning their community golf course into a shooting range, where their children collected slugs left behind.[20] The range is an ideal metaphor for the cultural battleground between militarism and maternalism, male science and female values that Los Alamos had become.

There was something psychologically pleasing about having women at Los Alamos who were not scientists. The maternity associated with them allowed the makers of the bomb to feel as if they were creating a benevolent tool, not a weapon of human destruction. (Then, too, men played psychological games with themselves by referring to their bombs as their "babies"—a perverse and reverse parthenogenesis of sorts.) Women delivered eighty babies during the first year at the site and about ten a month thereafter; from 1943 until the end of the war the population doubled at Los Alamos every ninety days. The baby boom wasn't surprising, given that the average age of residents was twenty-seven and the price of the hospital stay was a dollar a day, the estimated cost of food. Laura Fermi joked that it was a deal too good to refuse. Again, the image defies what we thought we knew about this scientific community.[21]

Los Alamos officially did not exist on a map. Residents referred to it simply as "the Hill," but to outsiders, it may as well have been the Bermuda Triangle. From almost any angle the surrounding mountains were breathtaking, the barbed wire and guards surrounding the site less so. Oppenheimer hoped the views would liberate the minds of his scientists; Groves thought the wooded mesa ideal for security purposes. The nearest neighbors, the Pueblo of San Ildefonso and Santa Clara, lived fifteen and twenty miles away, respectively; the next major town, Santa Fe, was more than thirty-five miles east by unpaved road. The buildings of a vacated prep school seemed sufficient accommodation for the handful of scientists and support staff first anticipated to move there in 1943. By the end of 1944, "PO Box 1663, Santa Fe" was the official postal address for nearly fifty-seven hundred residents.[22]

Mothers baby-proofed their porches with chicken wire, and the proliferation of laundry lines and roaming pets gave the place a slummy feel.[23] There was no mailman, no pavement, no bank, no milkman, no newspaper delivery, no laundry service (though the mothers of babies eventually instituted one). In regimental military style the place "ran on bells"—dinner at six, colloquiums at seven. Water shortages were so common that the army issued bulletins urging residents to soap up before entering the shower. The grounds between buildings were bone dry, and dust penetrated the cracks of apartment walls, creating residual film on living-room floors. Lois Bradbury hated that her "tenement" had paper-thin walls and windows that looked right into others in almost every direction. Families received army-issue furniture and monstrosities for stoves: the infamous coal-burning Black Beauties, which left homes so hot and sooty that women exchanged them for hot plates. One look at the furnace under the stairs of her second-floor apartment and Eleanor Jette's first mission was to find enough rope to tie a fire escape from her living room window.[24] Families of scientists lived in houses, while workmen settled in trailers scrunched together in repeating patterns of six to eight. Dormitories were scattered throughout. Single people occupied "Pacific huts" with long rows of cots, while couples got more private amenities in quickly erected apartments. The families of important scientists occupied Bathtub Row, the only accommodation furnished with the described luxury.

A distinctive sense of caste ruled Los Alamos. Ruth Marshak explained that "lines were drawn principally, not on wealth, family, or even age, but on the position one's husband held in the laboratory." Generally, physicists trumped chemists, technicians trumped computers, theoreticians trumped experimentalists, and more often than not this translated to men trumping women, when women figured at all. And yet one could also argue that Los Alamos felt hierarchical and egalitarian at the same time. The physical displacement of all involved—rich and poor, physicist and lab tech, man and woman—leveled the playing field somewhat. Where else could a world-class physicist like Otto Frisch turn thespian in an amateur production at the movie house and don the traditional dress of a Pueblo woman for the part? Such was the class, gender, and cultural crossover that occurred on the mesa; despite the military rigidity, the strangeness of situation and place occasionally made possible the transformation of self and the transgression of gender. People enjoyed their mixed drinks and mixed company in this oddly permissive society. At the playing fields, square dances, tennis courts, ski runs, and horse trails, division leaders cavorted freely with the men and women in their charge. University of Wisconsin masters student Joan Hinton served as assistant to Fermi in the Tech Area, but also joined him on the slopes and played violin in a quartet with senior scientists Frisch and Teller.[25]

In some ways Oppenheimer was responsible for the democratized feel of the place; no complaint was too minor, whether lodged by Nobel laureate or Indian domestic worker. In the Tech Area the prevailing attire was a casual pair of pants and buttoned-down plaid shirt; no one bothered to shine his or her shoes. Scientists shed their titles, and excepting Elinor Pulitzer, daughter of the famed newspaper magnate and bride of the Los Alamos medical director, women replaced hats, gloves, and high heels with bandanas and moccasins and left their nails and skin "shiningly natural," like those of their Pueblo domestic workers.[26]

Nevertheless, while social protocol was not always observed, there was some effort to separate the physical spaces of science and domesticity. The Tech Area lay beyond a wire fence, accessible only to scientists with white-badge clearance. Charlotte Serber, wife of Theoretical Physics Division group leader Robert Serber, had more access than most as the "scientific librarian," but other wives never saw inside the Tech Area unless they were specialists, computers, or support staff—and their clearance was low level. One wife came to begrudge everything associated with the Tech Area—"a great pit which swallowed our scientist husbands out of sight, almost out of our lives." Eleanor Jette likened it to the nest surrounding an insatiable ant queen; the bomb, like the post-war bombshell, was a female competing for her husband's attention. One newlywed wondered why she was no longer part of her husband's scientific world: "I had felt a sudden sense of isolation and anger when I discovered that Leon either wouldn't or couldn't answer my questions. . . . I felt uncomfortable, sensing a new distance between us."[27] Women unfit for the sacrifice of spousal attention bowed out, some divorcing or committing suicide. Men worried that science wives were becoming "depressed, quarrelsome, and gossipy" and needed something more to do. Psychologists advised Oppenheimer to get them involved in work situations, perhaps even to pay them so that they had tangible evidence of their usefulness.[28]

Wives of some of the top men lined up at the chance. Laura Fermi began working at the Los Alamos hospital, and Mici Teller offered her-self up to the militaristic machine, becoming a computer in the T-5 group of the Theoretical Physics Division. Hugh Richards discovered that inexperienced women were easy to train. After teaching a WAC photo emulsion techniques, he trained his secretary to do them too.[29] Frances Wilson Kurath, wife of physicist Dieter Kurath, already held a math degree when she joined T-5, but most science wives had no rel-evant experience before being called into service. Mary Frankel had a PhD and supervised the women with exacting standards. "She was fussy about decimals and regarded our mistakes as appalling," one wife re-membered. Yet under her, women computed the path of the shock wave from the fissile core of the bomb.

"*She* was the multiplier, and *she* was the adder, and this one cubed, and we had index cards, and all *she* did was cube this number and send it to the next one," Richard Feynman observed. But the nameless, face-less female assembly line turned out to be more efficient than the tem-peramental IBM machines he was called in to repair.[30] "Computers" op-erated on the technical side of things, yet received the pay of typists in an office. Most didn't complain, since they worked their required "three-eighths time" for the sole purpose of moving up on the housecleaning schedule. In May 1943, when the Security Office approved the busing in of Pueblo women from San Ildefonso to clean the homes of *work-ing* Los Alamos wives, women signed up for paid posts they had shown little interest in before. Technically, most women worked part-time, yet the reality was that secretaries, technicians, computers, draftsmen, and switchboard operators clocked regular seventy-hour weeks.

The employment office created a priority chart: ill or pregnant women got Native American domestic workers first, followed by full-time working wives, part-time working wives with children, nonworking wives with children, part-time working wives without children, and nonworking childless wives last. The pecking order was clear, but women complained anyway. Who gets more domestic help: a woman working two-thirds time with school-aged girls at home, or a mother working half-time with a rambunctious toddler boy? Who had the worse cross to bear? No one considered that the largest burden was perhaps on the overscheduled domestic worker who had her own children to feed.[31]

The all-consuming domesticity of science wives made their experience of Los Alamos different from that of single WACs and lower-level scientists and technicians who were not allowed to bring their families to the Hill. Many single women were swept into the social scene in the bachelor dorms, which military administrators feared had become cesspools of debauchery. The first men and women who occupied these dormitories had shared bathrooms and bunk beds until army officials restricted them to segregated quarters. "The street was a natural divider between male and female domains, but it was not a boundary," recalled army private Eleanor Stone. Men who wanted to date her friends went to the back of the barracks and quietly knocked to be let in. The young physicist Richard Feynman pleaded with the town council to let adults be adults and fraternize as they saw fit, but most restrictions remained, since they ostensibly curbed the spread of sexual disease. WACs who got pregnant were immediately discharged and shipped out; there was no place for such women within the social nexus of the Hill.[32]

Nevertheless, single women at Los Alamos confessed that, underneath the formalities, they had active sexual lives. There were about thirty GIs competing for every WAC, and women scientists were scarcer yet. Male scientists lamented the glut of datable women on the mesa; one attractive twenty-three-year-old received three marriage proposals while making milkshakes in the Tech Area PX. Hugh Richards, a new physics PhD, complained that the only single coeds of interest were nurses and Oppie's secretary, who quickly married another scientist. It was his great fortune that the head of the Metallurgy Division brought out a new secretary, a former science librarian, whom he married and with whom he had a son before they left Los Alamos.[33]

Richards was part of the Trinity Division and often left his family to set up cameras and signal leads near ground zero. The men in his group didn't operate jets on which they could paint likenesses of their spouses or favored starlets; they settled for painting a wife's name on a balloon tethered to a catcher camera several hundred meters from the projected Trinity explosion. The culture of the project, especially during its intense final phases, dictated science first, family later; and yet many young women relinquished careers and modest salaries to marry and settle down there in the brief period before the war ended. Dorothy McKibbin hosted at least twelve wedding ceremonies for scientists at her home in Santa Fe, and many others took place on the mesa between women and soon-to-be absentee husbands.[34]

In allowing their husbands to tend to the business of war, Los Alamos women became mythologized "wives as war martyrs." An image of them pining for their husbands' safe return—be they foot soldiers or explosives experts—perpetuated the fiction of war as a cleansing return to values, a necessary taking of lives to reinstate the wholesomeness of home. Eleanor Jette fancied herself in the image of this brand of heroine and complained little when her husband insisted that "no woman who can cook like you do has any business working in the Tech Area. . . . I've got enough problems without coming home to a cold dinner." Jette agreed that she did important supportive work at home, seeing her keeping of hot meals on the table as comparable to war nursing, rationing, or entertaining the troops; she kept morale up by assuring men that she hadn't forgotten her place in the social and sexual order: "My man is doing a job over there in the Tech Area and I'm here to take care of him."[35]

Although they didn't transgress the domestic sphere, some wives expanded it to include social and community functions for which their scientist husbands had no time. Los Alamos wives became active members of the town council, handling water shortages and traffic violations. Alice Kimball Smith, wife of the Metallurgy Division head, recalled the chronic delays in running the meetings, as most of the members drifted in after doing dishes or putting babies down. Teaching had long been the preserve of American women, and during the war female instructors replaced many of the male professors who vacated classrooms for the lab or battlefield. At Los Alamos, too, wives taught in the elementary and high schools. Ruth Marshak doubled as a third-grade teacher when not staffing the Housing Office; Jane Wilson taught high school English, and Alice Kimball taught history. Ironically, the subjects in which the high school remained deficient were the physical sciences their husbands advanced.[36]

Under such strange living conditions, the community might have lost all sense of order had women not taken charge. While their husbands worked until very late at night, self-sufficient women organized an increasingly separate social sphere in which male status and authority mattered little. While the military posted official bulletins, women transmitted information pertinent to their day-to-day lives through a grapevine they developed themselves. One historian described their social world as resembling "the sororities of college years . . . with big sisters to take the new 'girls' under their wings." In the Tech Area men organized themselves in a pecking order based on former individualist pursuits. But their wives' universe turned maternal and communitarian. As their husbands developed the technologies for a military arsenal, homemakers returned to precapitalist simplicity, bartering with Native American artisans for domestic goods.[37] When we take this underground society into account, in addition to all the ways in which women worked inside the male Tech Area, Los Alamos looks less like the bastion of masculinity it has been imagined to be.

The Woman Scientist as Oxymoron: The Distortion of Historical Memory

Mothers and wives entered scientific space, and the science for which Los Alamos was legendary was inextricable from the domestic lives of its scientists. And yet in hindsight women rarely noticed that they had bridged both worlds. They viewed themselves by definition as set apart from "the action"—the realms of scientific and historical import. Shortly after the war Jane Wilson and Charlotte Serber collected first-person essays from the women they knew at Los Alamos; each one described her role there as supportive, secondary, peripheral to important men. The title they chose for the collection was not *Helping Out and*

Taking Part, but *Standing by and Making Do*, an ironic decision in light of the editors' integral contributions in the science and community of Los Alamos; Serber, for one, had high-level clearance and division-head status as Oppenheimer's trusted confidante. Yet in her essays she grouped all women as helpmeets and passive bystanders. The collective ethos that gave women the strength to endure the conditions of Los Alamos diminished their stature in history books that paid greater homage to individual men.

Wives didn't acknowledge their import, but their daughters were cognizant of their different generational perspective. Karan McKibben, Gaby Peierls Gross, Katherine Manley, Wendy Teller, and other daughters of Los Alamos scientists confirmed that the experience shaped their views of science and womanhood. Some were inspired to become scientists themselves, while others took on the ethical burdens of their parents and spoke out against nuclear weapons as adults. While all eyes were on the great men of Los Alamos, Nella Fermi turned her attention to the women having babies. After the war she pursued a PhD in educational psychology, and rather than focus on the great men of the University of Chicago faculty, she wrote a dissertation on fertility patterns among their wives.[38]

It would be naive to think her father's stature and her mother's domesticity had nothing to do with her decisions. As a high school art teacher in Chicago, she couldn't help but see the personality types of her childhood in her students: "No girl would get away with the kind of behavior Bill shows in my class," she wrote in a pedagogical piece called "Sugar and Spice." "He is aggressive, even rude at times, and sprawls all over the chair as if he owned the world (maybe he does). But he has good ideas. I picture him as he may be twenty or thirty years from now—respected, listened to, a brilliant man who had made his work in some intellectual field or another. Eccentric, yes—egoistic, yes, but forgiven this because of his indubitable achievement. . . . Could a girl be like that? I can't imagine one."[39]

Los Alamos left its mark on the daughters of scientists and eventually on all American women seeking scientific status in the postwar years. The cultural memory of Los Alamos morphed into a gendered parable and a cautionary tale, warning against females who trespassed into scientific realms. Such memoirs as Leona Marshall Libby's *The Uranium People* (1979) and Elizabeth Rona's *How It Came About* (1978) confirm women's participation in wartime nuclear science. But both authors agree that only their ability to assimilate into masculine space gave them license to write about themselves as participating scientists. Perhaps they assimilated too well, for few of their male colleagues thought them standouts worth mention in their own historical accounts. Glenn Seaborg provided a rare glimpse of women technicians in his voluminous account of the Met Lab, but they appeared often as participants in "lab romances" and have been difficult to trace once they took married names.[40]

Women themselves have been responsible for neglecting their own histories as they venerate the "great men" of science. In her memoir *Atoms in the Family* (1953), Laura Fermi barely mentions Leona Marshall Libby, a woman integral to her husband's research, to the development of the bomb, and to Laura's own private life during those years. Misleadingly subtitling her book *My Life with Enrico Fermi*, Laura focuses on herself only sporadically during the early years when she studied physics and mathematics at the University of Rome, dropping her scientific identity into the shadow of Enrico once they marry in 1928.[41] From that point on, she claims ignorance about all things technical; Enrico refused to "talk shop" in her presence because, as she explains, she was "a slow thinker" who failed to "grasp scientific matters easily." At get-togethers with science couples, she reports, men embarrassed their wives by displaying the latter's deficiencies, Laura's included: "I gradually developed an overconsciousness of my ignorance, of the worthlessness of my own opinions." By the time Libby met Laura Fermi in the 1940s, her transformation was complete: "She never asked a question about the physical world, never showed the least curiosity about the Manhattan District," Libby remembered. "Her life was fulfilled by children, home, women friends, clothes, food, and domesticity."[42]

Much as Anne Roe depicted her prototypical science heroes in boyhood, as noted in the introduction to Part 2, Laura Fermi portrayed young Enrico as self-sufficient, introspective, and gadgeteering—an "individualist" who "value[d] his independence above all." For all their intimacy as husband and wife, she wrote of her husband in worshipful terms: "In matters of faith the pope is infallible. In quantum theory Fermi is infallible. Ergo Fermi is the pope." She cowered in her husband's divine presence, an attitude made manifest in the photographs she chose to accompany her text. Enrico is always prominent, shown in the company of other famous scientists (all male), while she appears shrouded, even in her own wedding photo. As in the text, she dwells in the background of more important members of the male world.[43]

Helpmeets and historians idolized and idealized the men of the Manhattan Project, but of course such men were also not shy about seeing themselves as heroic figures, often leaving grist for others to add to their legends. Taken together, these voices have perpetuated an image of the Manhattan Project as a male rite of passage, a maker of masculine heroes. "Not since classical times had there been such a collection of talent gathered in one place," project scientist Nicholas Metropolis mythologized in 1992.[44] It's not surprising that he couched his experience as a romantic narrative, for this is how Oppenheimer pitched the project to the young scientists he recruited back in 1942. Robert Wilson recalled the hard sell:

> The projected laboratory as described by Oppenheimer sounded romantic—and it was romantic. Everything to do with it was to be clothed in deepest secrecy. We were all to join the army and then disappear to a mountain-top laboratory in New Mexico—Los Alamos. It sounded especially romantic to me for I had just finished reading *The Magic Mountain* by Thomas Mann. I almost expected, empathetically, to come down with tuberculosis, certainly expected to explore the philosophical significance of concepts of time and space and freedom and fascism in intense conversations with an Italian philosopher in a snowstorm. And I did, too! Not with Mann's fictional Settembrini, but rather with a real, live, breathing Enrico Fermi. And Fermi was better than Settembrini.[45]

Mary Palevsky, the daughter of low-level scientists at Los Alamos, continues to marvel at the extent to which such hero worship has been self-sustaining, even in the twenty-first century. "In recent obituaries of project scientists," she notes, "the closer they were to Oppenheimer and the bomb, the more the scientific prowess that accrues." The *New York Times* heralded her own father as "an A-bomb Developer," though in no way was he ever a principal scientist in the project. In this she sensed the cultural significance Americans still attribute to the bomb itself; they understand it and its creators almost purely in terms of power and prestige. Perhaps women would have ridden the same coattails had they established their place in project lore early on. Ellen Weaver, a former Oak Ridge scientist, recognized that they missed their chance: "When I read the personal reminiscences of the men who were nuclear pioneers,

I'm struck by the importance they place on their friends and associates, and on the often intense interaction among them. . . . By and large, women did not share in that give-and-take."[46]

The historical invisibility of women scientists of the project can be attributed also to our conception of science as a masculine, solitary pursuit. We continue to venerate the lone maverick, male by definition, rather than the collective experience for which evidence of a female presence is greater. This lone maverick is also endearingly eccentric, a trait that has been imagined as unique to male genius. Physicists who isolated electrons and built cyclotrons have been described over the years as "baffled by can openers." Eleanor Jette described physicist Niels Bohr the same way biographers have the mythical Pierre Curie: as an absentminded professor, too stuck in his head to notice worldly things. Wrapped up in thought as he walked the Paris streets, unsuspecting Pierre was killed by a horse and carriage. According to Jette, Bohr almost suffered the same fate in front of the Tech Area of Los Alamos. A historian took it on the authority of a Los Alamos chef that world-class scientists walked the grounds aimlessly at 4:00 a.m., perpetually caught up in theoretical problems. Women rarely have been depicted as exhibiting such single-minded intensity. Even when Marie Curie has been so described, she has never simultaneously been a likable, larger-than-life personality. For her, there has always been a trade-off; her scientific mind has relegated her to a life of serious and quiet contemplation.[47]

Biographers have never criticized Albert Einstein's preference for unkempt hair and have viewed Leo Szilard's refusal to flush toilets with amusement. Richard Feynman's safecracking, codebreaking, bongo beating, and acts of "active irresponsibility" have made him famous; one biographer referred to him admiringly as "all genius and all buffoon." Although images of nonconformity might have liberated all scientists from stereotypical molds, invariably they have stigmatized one sex while elevating another. Physical chemist Nathalie Michel Goldowski was a jovial George Gamow type in heels—literally. While the project scientists around her wore flannel, this heir to Russian aristocracy donned the latest in Paris fashions. Plutonium production would have ceased without the aluminum coating she developed for the uranium slugs in the Hanford reactors, but she has not been passed down to posterity as an innovator or even as an intriguing figure comparable to her male counterparts.[48]

Is it too late to move women scientists integrally into project history? Oral and written accounts are limited, and official documents shed little light. Supervisors recorded discoveries in laboratory annals but rarely included the names of the women who made them. Trying to collect the names of women who worked for OSRD projects was a trial for historian Margaret Rossiter, who found that most were anonymous in official project histories. Even a report she retrieved specifically on women's contributions to the OSRD didn't include names of individuals. Because so many women were recruited informally, they left less of a paper trail than did men, for whom recruitment trips were made. No amount of official paperwork would ever indicate the significance of Elizabeth Graves or Klara Dan von Neumann, wives of prominent Los Alamos men who, once in New Mexico, were pressed into service and became indispensable contributors to the project. Elizabeth Rona, who worked with Meitner in Berlin and Curie in Paris, didn't have the working papers to be acknowledged formally as a project participant; we only know her story because she bothered to write it down.[49]

One can scan the lists of project brass for a long time before finding a woman in the ranks. Mina Rees and Gladys Amelia Anslow were perhaps the highest up, but romantic images of the project diminish their significance, since they were largely administrators, not scientists in the trenches. As one pans outward from the inner core of project coordinators and down past the division heads, more women come into focus. Some of Fermi's assistants were involved in the nuclear science of the bomb at several sites. Kay Way calculated the flux of neutrons in CP-1, the first atomic pile at Chicago, and contributed to understanding fission-product decay (the Way-Wigner formula), which aided in the designing of reactors at Hanford and Oak Ridge. Unfortunately, her absence at Trinity has sealed her anonymity to those who insist that everybody who was anybody was there. Historians have remained fixated on the epicenter of both the explosion and formalized channels of power.

The Los Alamos Historical Society maintains ledgers of project employees that reveal women participants (when their names are not ambiguously gendered). And yet much of the archived literature only amplifies the extent to which women scientists were made invisible. Memoranda for Los Alamos residents, for example, refer to scientists always as men and spouses always as wives. Such documents fuel the presumption that single scientists such as Joan Hinton didn't exist, and yet hers is a compelling story. She came to Los Alamos in 1944 to test assemblies of enriched uranium and plutonium and worked in a science building away from the others—down a cliff—for fear that her reactor might go critical. The precaution proved warranted when her colleague Harry Daghlian suffered exposure to lethal levels of radiation in the lab. During an experiment to measure the critical mass of uranium-bomb assemblies, he dropped a paraffin block onto the reactor's active plutonium sphere and was irradiated through his hands, which turned gangrenous over the next three weeks. Hinton witnessed his slow and torturous death, but she was more than a mere witness to such events. Physics is a field that venerates youth, and Hinton was the youngest scientist working on the bomb. Her participation in the science was direct, though after the war it weighed too much on her conscience to acknowledge it.[50]

In the 1990s physicists Ruth Howes and Caroline Herzenberg retrieved information on three hundred women who did scientific work at project sites. Separating the women's contributions from men's was a complex process that required reading between the lines drawn by historians and the women themselves. Since none of these women were part of the highest-ranking leadership, many internalized their subordinate status decades later, attributing their own breakthroughs in the lab to the men who supervised them. Most who worked in technical areas were scattered and often isolated from other women, and hence developed no sense of collective consciousness. At Oak Ridge Ellen Weaver saw women in the locker room changing into protective clothing, but she offered no gestures of friendship. "My buddies were all men," she recalled. "I'm ashamed to say that I occasionally heard men at Oak Ridge discuss female scientists in insulting terms . . . but that I did not rise to their defense. I remember no sense of camaraderie among women; indeed, we viewed other women with some suspicion."[51]

In the project hierarchy, the theoretical physicist was king, but Howes and Herzenberg were hard pressed to find corresponding queens. Women such as Naomi Livesay French and Lillian Carson were in the Theoretical Division at Los Alamos, but not as physicists. Chien-Shiung Wu, Maria Goeppert Mayer, Helen Jupnik, Miriam Gilbert, Lydia Savedoff, and Jane Hamilton Hall were physicists at Columbia, Princeton, Clinton Engineering Works, and Hanford, and thus have been portrayed in the historical literature as peripheral to the core figures at Los Alamos, when they have been discussed at all. Sue Norton, Dorothy Wallace, and Lyda Speck were physicists in the Met Lab of Los Alamos, but worked in the Health Division or in technician positions where their anonymity was assured. Joan Hinton and Elizabeth Graves were experimental physicists, and thus not part of the "theoretical" elite. Historians have also disregarded such biophysicists as Elda Anderson and Edith Quimby, whose pioneering research on the effects of radio-activity was crucial to the project but was minimized in the afterglow of the bomb. Mary Argo was a theoretical physicist at Los Alamos, yet her obscurity was likely the result of working on the fusion bomb under Teller rather than on the fission bomb with Fermi. Columbia physicist Elky Shazkin was eliminated from the project altogether; nepotism policies forced her to watch from the sidelines while her husband participated in a formal capacity.[52]

Howes and Herzenberg found more women chemists than physicists and more "technicians," "lab workers," and "computers," than scientists and group leaders, but titles don't indicate all there is to know about women's work on the project. Mary Miller would likely have received more credit as a group leader at Los Alamos had her PhD in physical chemistry not been obscured by her low military rank as a WAC. Although Mary Argo was the only woman formally cleared to witness Trinity, we know that women worked in the Explosives Division of the project, at least twenty directly on engineering the bomb. Frances Dunne, a senior aircraft mechanic, had the dexterity and small hands to run the explosives testing for Fat Man and Little Boy: only she could reach inside the cavity of the explosive to assemble the trigger to run the trials. Dunne's work was experimental, not theoretical, yet it is inaccurate to characterize men as the thinkers and women as the doers of the project. When all ranks of physicists are considered, women theorists outnumbered experimentalists four to one.[53]

Retrieving women of the project would require that military historians become social historians who privilege sources revealing the project from the bottom up and outward in. Historians must look into university labs as well as production sites at Hanford and Oak Ridge, not only at Los Alamos; they must also examine personal and nonofficial writings and search for oral testimony. They have to look in nooks that elude traditional historians of science and redefine notions of "work" on the bomb. When the conception of the project is expanded to include mathematics, medical research, and chemistry, as well as hands-on and multitiered clinical, bench, and computational work, women will surface as significant participants.

Notes

1. Ruth Marshak, "Secret City," in *Standing by and Making Do: Women of Wartime Los Alamos*, ed. Jane S. Wilson and Charlotte Serber (Los Alamos, NM: Los Alamos Historical Society, 1988), 1.

2. Eleanor Jette, *Inside Box 1663* (Los Alamos: Los Alamos Historical Society, 1977), 45–46.

3. Oppenheimer quoted in Ruth Howes and Caroline Herzenberg, *Their Day in the Sun: Women of the Manhattan Project* (Philadelphia: Temple University Press, 1999), 150.

4. Elizabeth Rona, *How It Came About: Radioactivity, Nuclear Physics, and Atomic Energy* (Oak Ridge: Oak Ridge Associated Universities, 1978), 53–57.

5. Jette, *Inside Box 1663*, 5–23, 33.

6. Elsie McMillan, "Outside the Inner Fence," in *Reminiscences of Los Alamos, 1943–1945*, ed. Lawrence Badash, Joseph O. Hirschfelder, and Herbert P. Broida (Holland: D. Reidel, 1980), 46–47.

7. Marshak, "Secret City," 1, 6.

8. Charlotte Serber, "Labor Pains," in *Standing by*, 60–62; Jane Wilson, "Not Quite Eden," in *Standing by*, 45–46, 51–52, 54; Marshak, "Secret City," 3–5; Laura Fermi, "Fermi's Path to Los Alamos," in *Reminiscences of Los Alamos*, 90–96; Leona Marshall Libby, *The Uranium People* (New York: Charles Scribners' Sons, 1979), 26–27.

9. Evelyn Fox Keller notes the perception of science as antithetical to Eros and thus tied to sexual restraint. See *Reflections on Gender and Science* (New Haven: Yale University Press), 21–32, 75–94; Peter Bacon Hales, *Atomic Spaces: Living on the Manhattan Project* (Urbana: University of Illinois Press, 1997), 219–21.

10. Ruth Howes and Caroline Herzenberg, "Women in Weapons Development: The Manhattan Project," in *Women and the Use of Military Force*, ed. Ruth H. Howes and Michael R. Stevenson (Boulder, CO: Lynne Rienner, 1993), 95–109; *Their Day in the Sun*, 13–14, 168–70; Rachel Fermi and Esther Samra, eds., *Picturing the Bomb: Photographs from the Secret World of the Manhattan Project* (New York: Harry N. Abrams, 1995), 16; Hales, *Atomic Spaces*, 218–19.

11. Ida Noddack, "Über das Element 93," *Zeitschrift für Angewandte Chemie* 37 (1934): 653; Helge Kragh, *Quantum Generations: A History of Physics in the Twentieth Century* (Princeton: Princeton University Press, 1999), 257–58; L. M. Jones, "Intellectual Contributions of Women to Physics," in *Women of Science: Righting the Record*, ed. G. Kass Simon and Patricia Farnes (Bloomington: Indiana University Press, 1990), 193–95; Ruth Lewin Sime, *Lise Meitner: A Life in Physics* (Berkeley: University of California Press, 1996), 305–6; Londa Schiebinger, *Has Feminism Changed Science?* (Cambridge, MA: Harvard University Press, 1999), 167.

12. Dorothy McKibbin, "109 East Palace," in *Standing by*, 20–27; Jennet Conant, *109 East Palace: Robert Oppenheimer and the Secret City of Los Alamos* (New York: Simon and Schuster, 2005); Bernice Brode, "Tales of Los Alamos," in *Reminiscences of Los Alamos*, 147.

13. Libby, *Uranium People*, ix, 4, 130.

14. Libby, *Uranium People*, 126.

15. Herbert L. Anderson, "Assisting Fermi," in *All in Our Time: Reminiscences of Twelve Nuclear Pioneers*, ed. Jane Wilson (Chicago: The Bulletin of the Atomic Scientists, 1974), 87, 96; Libby, *Uranium People*, 26–27, 129.

16. Libby, *Uranium People*, 1, 234–36.

17. Libby, *Uranium People*, 155–56.

18. Libby, *Uranium People*, 164–65, 183–84.

19. Howes and Herzenberg, "Women in Weapons Development," 99, 104, 109; *Their Day in the Sun*, 49, 125–27, 193.

20. Mary Palevsky, *Atomic Fragments: A Daughter's Questions* (Berkeley: University of California Press, 2000), 101–2; Jette, *Inside Box 1663*, 99.

21. Brian Easlea, *Fathering the Unthinkable: Masculinity, Scientists, and the Nuclear Arms Race* (London: Pluto, 1983); Carol Cohn, "Slick'ems, Glick'ems, Christmas Trees, and Cookie Cutters: Nuclear Language and How We Learned to Pat the Bomb," *Bulletin of Atomic Scientists* 43 (June 1987): 20; Shirley B. Barnett, "Operation Los Alamos," in *Standing by*, 92–93; Brode, "Tales of Los Alamos," 138; Marshak, "Secret City," 15–16, 19; Fermi and Samra, *Picturing the Bomb*, 17; Katrina R. Mason, *Children of Los Alamos* (New York: Twayne, 1995), 9; Fermi, "The Fermis' Path to Los Alamos," 94; Howes and Herzenberg, *Their Day in the Sun*, 154; Ferenc Morton Szasz, *The Day the Sun Rose Twice: The Story of the Trinity Site Nuclear Explosion, July 16, 1945* (Albuquerque: University of New Mexico Press, 1984), 18.

22. Mason, *Children of Los Alamos*, ix–x; Conant, *109 East Palace*, 39–44.

23. Kathleen Mark, "A Roof Over Our Heads," in *Standing by*, 29–37; Marshak, "Secret City," 5.

24. Jane Wilson, "Not Quite Eden," in *Standing by*, 47–49; Mark, "A Roof Over Our Heads," 39–40; Jette, *Inside Box 1663*, 17, 33; Hales, *Atomic Spaces*, 212.

25. Jean Bacher, "Fresh Air and Alcohol," in *Standing by*, 103–15; Howes and Herzenberg, *Their Day in the Sun*, 55.

26. Marshak, "Secret City," 9; Serber, "Labor Pains," 58; Charlie Masters, "Going Native," in *Standing by*, 119–20; Jette, *Inside Box 1663*, 48; Conant, *109 East Palace*, 123–24.

27. Serber, "Labor Pains," 65; Marshak, "Secret City," 10–11; Jette, *Inside Box 1663*, 29, 42; Phyllis K. Fisher, *Los Alamos Experience* (Tokyo: Japan, 1985), 26 (Fisher cited in Hales, *Atomic Spaces*, 211).

28. Conant, *109 East Palace*, 142–45; Marshak, "Secret City," 10–11.

29. Hugh Richards, "The Making of the Bomb: A Personal Perspective," in *Behind Tall Fences: Stories and Experiences about Los Alamos at Its Beginning* (Los Alamos: Los Alamos Historical Society, 1996), 123.

30. Brode, "Tales of Los Alamos," 147–48; Libby, *Uranium People*, 236; Howes and Herzenberg, "Women in Weapons Development," 105–6; *Their Day in the Sun*, 99–100; Richard P. Feynman, "Los Alamos from Below," in *Reminiscences of Los Alamos*, 125.

31. Alice Kimball Smith, "Law and Order," in *Standing by*, 79; Serber, "Labor Pains," 57–68.

32. Feynman, "Los Alamos from Below," 111; Robert Wilson, "A Recruit for Los Alamos," in *All in Our Time*, 160; Eleanor Stone Roensch, *Life Within Limits* (Los Alamos: Los Alamos Historical Society, 1993), 13–14, 17; Jette, *Inside Box 1663*, 98.

33. Bacher, "Fresh Air and Alcohol," 112–13; John Manley, "A New Laboratory Is Born," in *Reminiscences of Los Alamos*, 29; Roensch, *Life Within Limits*, 17; Richards, "The Making of the Bomb," 122–27.

34. Richards, "The Making of the Bomb," 127–134.

35. Jette, *Inside Box 1663*, 29, 67, 69.

36. Smith, "Law and Order," 73–77; Mason, *Children of Los Alamos*, 79; Brode, "Tales of Los Alamos," 151.

37. Hales, *Atomic Spaces*, 207, 214–15.

38. Karan McKibben, "Behind Tall Fences," in *Behind Tall Fences*, 179; Mason, *Children of Los Alamos*, 79.

39. Nella Fermi Weiner, "Sugar and Spice," *School Review* 81, no. 1 (1972): 96–99.

40. Glenn T. Seaborg, *History of Met Lab Section C-I: April 1942 to April 1943; May 1943 to April 1944; May 1944 to April 1945*, vols. 1–3, Lawrence Berkeley Publication PUB 112, University of California, Berkeley (February 1977, May 1978, May 1979); ed. Ronald L. Kathren, Jerry B. Gough, and Gary T. Benefiel, *The Plutonium Story: The Journals of Professor Glenn T. Seaborg, 1939–1946* (Columbus, OH: Battelle Press, 1994).

41. Marriage created a similar transformation in Kitty Puening, a biologist, when she married Robert Oppenheimer in 1940. From thereafter she grudgingly played the role of faculty wife and hostess. See Conant, *109 East Palace*, 179–89.

42. Laura Fermi, *Atoms in the Family: My Life with Enrico Fermi* (Chicago: University of Chicago Press, 1954), 60–66, 156; Libby, *Uranium People*, 23–24.

43. Fermi, *Atoms in the Family*, 16–19, 47, 59, 212.

44. Nicholas Metropolis, "Random Reminiscences," in *Behind Tall Fences*, 69.

45. Wilson, "A Recruit for Los Alamos," 143.

46. Palevsky, *Atomic Fragments*, 223; Weaver quoted in Howes and Herzenberg, *Their Day in the Sun*, viii.

47. Jette, *Inside Box 1663*, 77–78; Szasz, *The Day the Sun Rose Twice*, 21–22; Schiebinger, *Has Feminism Changed Science?* 73.

48. Feynman, "Los Alamos from Below," 112–17, 129; Richard P. Feynman, *"Surely You're Joking, Mr. Feynman!"* ed. Edward Hutchings (New York: W. W. Norton, 1985); John Gribbin and Mary Gribbin, *Richard Feynman: A Life in Science* (New York: Dutton, 1997); George Greenstein, *Portraits of Discovery: Profiles in Scientific Genius* (New York: John Wiley and Sons, 1998), 122; Libby, *Uranium People*, 63–65; Schiebinger, *Has Feminism Changed Science?* 76; Howes and Herzenberg, "Women in Weapons Development," 102–3.

49. Margaret Rossiter, *Women Scientists in America: Before Affirmative Action, 1940–1972* (Baltimore: Johns Hopkins University Press, 1995), 4–9.

50. Restricted memorandum published as appendix in Jette, *Inside Box 1663*, 128; Catherine Rampell, "The Atom Spy That Got Away," NBC News, August 13, 2004, http://www.msnbc.msn.com/id/5653644; "Facing Life," *Time*, August 9, 1954, http://www.time.com/time/magazine/article/0,9171,936255-2,00.html; Howes and Herzenberg, *Their Day in the Sun*, 51–56; Libby, *Uranium People*, 202.

51. Weaver quoted in Howes and Herzenberg, *Their Day in the Sun*, vii, 198.

52. Howes and Herzenberg, "Women in Weapons Development," 106; *Their Day in the Sun*, 50–51, 58–59, Appendix: "Female Scientific and Technical Workers in the Manhattan Project," 203–18.

53. Howes and Herzenberg, *Their Day in the Sun*, 14; "Women in Weapons Development," 95–109.

1. The Marie Curie tour, 1921. From left: Missy Meloney, Curie's daughter Irene, Marie Curie, and Curie's daughter Eve.

2. Lillian Gilbreth and family in Nantucket, 1923.

3. Staff analyzing stellar photographs and computed data at the Harvard College Observatory, 1891. Standing, from left: Edward Pickering, Williamina Fleming.

4. Scientists outside Eckhart Hall at the University of Chicago to commemorate the fourth anniversary of the first nuclear chain reaction under Stagg Field. Leona Marshall is the only woman, and Leo Szilard is to her right. Enrico Fermi and Walter Zinn, her lab supervisors at the University of Chicago and Argonne, are the first two men in the front row, respectively.

5. A party at a home on Bathtub Row, Los Alamos, September 1946. Eleanor and Eric Jette mingle with Robert Oppenheimer, Betty Brixner, and Irma and Frank Walters.

6. Charlotte Serber (front row, center) and the female staff of the TA-1 technical library at Los Alamos.

7. Unidentified woman and daughter hanging laundry in the Quonset hut area of Los Alamos, 1945.

8. Jane and David Hall at work in front of an instrument panel in a Los Alamos laboratory, 1946.

9. Maria Goeppert Mayer with Enrico Fermi (center)
and colleagues, date unknown.

10. Barbara McClintock watering her corn plants at Cold
Spring Harbor, 1953.

11. Jane Goodall with chimpanzees in *National Geographic*, circa 1965.

12. Dian Fossey and gorillas in *National Geographic*, 1981.

13. Biruté Galdikas and orangutans in *National Geographic*, 1975.

14. Galdikas's son, Binti, and Princess the orangutan bathe together, 1980.

5 Maria Goeppert Mayer and Rosalind Franklin: The Politics of Partners and Prizes in the Heroic Age of Science

One may ask as a last question whether the fact that she was a woman had any influence on her career as a scientist. This does not appear to be the case. She possessed a unique combination of feminine charm and a keen analytical mind. She was happily married with a man of similar inclinations and tastes with whom she could collaborate very closely. Finally, she was sought out as a co-equal discussion partner and friend by such greats as Max Born, Eugene Wigner, Edward Teller, and Enrico Fermi. She was cast from a similar mold.

—Eulogy by physicists Willard Libby, Bernd T. Matthias, Lothar Nordheim, and Harold Urey upon the death of Maria Mayer, 1972[1]

Our efforts have been largely complementary, and one without the other would not have gone as far as in combination.

—Carl Cori on his work with wife and fellow Nobelist Gerty Cori, Nobel Banquet Speech, 1947

If she had had someone to talk to, chances are she would have gotten to DNA first; it was all there in her notes and photographs . . . in no sense was she in the club, anyone's club . . . Franklin was something of a permanent freak in their midst.

—Vivian Gornick, *Women in Science: Then and Now*, 2009

It was 1948 at Argonne Laboratories, just outside Chicago. For two years the theoretical physicist Maria Goeppert Mayer had been working with her colleague and friend Edward Teller on one of the most basic questions in the physical universe: the origins of elements, starting with the nature of neutronic matter after the Big Bang. Teller hated working alone; he had recruited Mayer because he knew no other person with equal mathematical skills and a tolerance for his musings. But to his irritation she had become sidetracked by the peripheral problem of "magic numbers." Back in 1933 Walter Elsasser had noted them but hadn't figured out their significance. Years later, when she took up the question again, Teller recalled her fascination:

In examining the facts our theory had to explain, Maria noticed that those nuclei that had either 2, 6, 14, 28, 50, 82, or 126 protons or neutrons were far more abundant than nuclei with not very different proton or neutron numbers. . . . If both the neutrons and protons were of a "magic" number, the isotope was particularly abundant.

That seemed like a detail to me, but Maria thought that the regular repetitious appearance of these abundances must have an interesting explanation in itself; whether it was connected with the origin of elements was not the issue. I persisted in disparaging her interest until finally she lost her temper.[2]

Teller left for Los Alamos, leaving Mayer to ponder the magic numbers on her own. Helium, argon, and xenon were all gases that didn't readily change into other elements, and they were highly stable, with magic numbers of electrons tightly bound to the nucleus. Abundance was a factor in these numbers, to be sure, but so were elements' spins, binding energies, and magnetic moments. Were beta-decay properties or quadropole moments also factors? Mayer was mulling over her numbers in her office with Enrico Fermi, when he got up to take a long-distance phone call. He turned in the doorway as he was leaving: "Is there any indication of spin-orbit coupling?" Mayer chewed on his suggestion, even though it ran counter to the commonly accepted "liquid drop" model of the atomic nucleus. Most scientists assumed that coupling was very weak, but Mayer had lived with the data so long that she could immediately see how much Fermi's suggestion explained. She scribbled equations madly. When Fermi returned she explained her epiphany in rapid fire: In the nucleus the intrinsic spin of every nucleon is strongly coupled to the angular momentum of its own orbit. The actual nucleus was that in which the protons and neutrons were at the level of lowest energy permitted by the exclusion principle. Fermi was skeptical and left her to her numerology.[3]

That night the children eagerly awaited their mother, but Mayer was too consumed to talk with them. She and her husband, chemist Joe Mayer, poured cocktails and went through the notes she had amassed on her magic numbers. Cigarette after cigarette burned as they plugged one element after the next into equations that confirmed she was on to something, and days later she returned to Argonne with the write-up of her shell-model theory of the nucleus. Now with more time to look at her ideas, carefully laid out, Fermi agreed that her theory was too elegant to be wrong and started teaching the model to graduate students as if it had long been accepted.[4] Fourteen years later that model won a Nobel Prize for Maria Mayer.

When Mayer was an instructor at Columbia, she had a colleague named Chien-Shiung Wu, an experimental physicist, who seemed well positioned to win a Nobel Prize of her own. She had come from China in 1936 to work with Ernest Lawrence at Berkeley and had become a leading expert on fission and beta decay by the time she submitted her PhD thesis in 1940. Administrators at Berkeley hesitated to offer her a position, not only because she was a woman, but also because of strong anti-Asian sentiment on the West Coast. She moved east to teach women at Smith College, just as her expertise in nuclear physics opened up unprecedented opportunities for research. She was offered replacement positions for faculty at Princeton and MIT who had left for war work, and in 1944 she joined Manhattan Project scientists, including Mayer, at Columbia. Mayer left for Chicago in 1946, but Wu stayed on and became a full-time, albeit untenured, faculty member in 1952.[5]

Junior colleagues moved up the ranks more quickly, but Wu never complained. Tsung-Dao Lee was fourteen years her junior and tenured when he walked into her office in 1956 to seek her expertise on weak interactions, for he wanted to prove that parity was not conserved in them. Wu was singularly able to produce experiments requiring conditions at absolute vacuum and near absolute zero, and she agreed to take up his problem. For the following six months she worked tirelessly in New York as well as Washington, since the equipment and staff she needed for low-temperature spin polarization was at the Bureau of Standards. Her experiments were elegant in their simplicity, beautiful in their definitiveness. Observation of beta particles given off by cobalt-60 made it clear that there was a preferred direction of emission; parity was not conserved for this weak interaction.[6]

Word of her findings spread quickly; she posed with Lee and Yang on the cover of the New York Times and was featured in Newsweek and Time. The Nobel Prize became a foregone conclusion ten months later, but she did not join her colleagues in Stockholm. By the time her group had written up results, others had closed in; apparently it was easier to differentiate the originators of the idea to disprove parity than to tease out the first and best experimentalist on the case. Physicist Noemie

Koller echoed the sentiments of many who believed that Wu deserved the prize and that sex discrimination was to blame. She "straightened up a big mess in physics quite elegantly," Koller concluded, but Nobels are not awarded for cleanup; they go to the conspicuous parties once all the dust has cleared.[7]

If there is truth to the lore of how Wu's mentor, Ernest Lawrence, came up with the cyclotron that won him his Nobel Prize (1939), it may provide insight into the dynamics that worked against her. Reportedly Lawrence sat in a Berkeley library reading a paper on accelerating charged particles when the idea of the cyclotron burst forth. He was so sure that he had hit the mother lode that he boasted he was going to be famous—and he wasn't wrong. The pure ideas that came to mind like a flash of light—the theory of relativity, nuclear fission—weren't supposed to come to the plodding and deliberate minds of women. Scientific virtuosity would not describe Marie Curie's years of burning off pitchblende or Wu's meticulous building of apparatus. Nobel Prizes went to the men who theorized the ideas women conscientiously confirmed at the bench.[8]

Of course none of this explains what happened to Lise Meitner, the theoretical physicist who did indeed come up with the concept of fission. She had been one-half of a thirty-year partnership with chemist Otto Hahn, but she was a Jew and forced to flee Berlin in 1938, leaving her uranium experiments behind for Hahn and his assistant, Fritz Strassman, to run. She settled in Stockholm and, from a distance, remained the "intellectual head" of the group, at least as Strassman described it. That winter Hahn sent her a letter about the results of his latest experiments. He had bombarded uranium with neutrons and produced barium fragments that he couldn't explain. Meitner figured that the barium was a product of an unknown process and stewed over the problem until she conceived what's known as fission. Hahn won the Nobel Prize for the discovery in 1944.

In dismissing Meitner's intellectual contributions to his Nobel-winning work, Hahn also dismissed her part in all the knowledge he had acquired to that point. Together they had discovered thorium in 1908 and protactinium in 1917, and together they had conducted studies on nuclear isomerism. When Hahn was conscripted into the German army during World War I, Meitner was left to isolate protactinium on her own, yet she listed him as first author when she reported the discovery. In the case of fission, he did not return the same courtesy, despite the fact that she had been running uranium experiments for four years before fleeing to Sweden. When peers raised the question of her Nobel-deservedness, Hahn called fission a problem of experimental chemistry, not theoretical physics. His scientific arguments were flimsy, but other factors probably contributed to his minimizing her work. Hahn was an iconic figure, a former war hero, whose successful science improved the collective psyche of Germans during World War II. To credit Meitner for his discovery not only would have tarnished his mythic status, but also would have necessitated confronting the persecution of a Jewish woman, thus opening a political Pandora's box. A recent biographer believes that Hahn and the German establishment thus colluded to keep the box closed.[9]

Nevertheless, the real rub for Meitner was not losing the Nobel, but gaining, in references to her thereafter, the label of Hahn's *Mitarbeiterin*, a German term whose meaning is closer to the English "assistant" or "subordinate" than to the "partner" she was. Marie Curie suffered a similar reduction in status in relation to Pierre, as did Rosalind Franklin in relation to her King's College colleague Maurice Wilkins. Whatever the realities of the relationships between these pairs of scientists, history has tended to obscure the contributions of women, who have rarely been seen as in charge. Collaborative science is the norm, yet Nobel's policy of rewarding a prize to no more than three individuals in any field continues to obscure the contributions that men—and especially women—make to discovery. The inability to tease out one scientist's contributions from another's isn't inherently sexist, and yet the idealization of certain individuals has made it possible to overlook those who have not had comparable titles or salaries and thus have worked invisibly. Historically, women have suffered such anonymity. As late as 1974 the Nobel Prize for the discovery of pulsars went to astrophysicist Anthony Hewish, though Jocelyn Bell, his graduate student, first noted the stellar radio source he later identified. Cosmologists wondered whether she too should have received the award, but Bell humbly dropped the issue. Students don't win Nobel Prizes, she insisted, though male graduate students have proved otherwise.[10]

Back at the turn of the twentieth century, biologist Nettie Stevens (1861–1912) proved that the x and y chromosomes in beetle sperm determined sex in a fertilized egg. Her observations were duly noted, but over decades and with mounting studies by male biologists, her pioneering insights have been omitted from textbooks. Stevens didn't begin studying science until she was thirty-five, and she died nine years after earning her PhD, well before establishing an esteemed reputation. Most of the credit for her work thus went to lab director E. B. Wilson and her graduate mentor, Thomas Hunt Morgan, who won a Nobel Prize for pioneering the field of genetics.[11]

This is not to say that men haven't been snubbed: the Nobel Committee passed over Oswald Avery and Erwin Chargaff before overlooking Rosalind Franklin in her same area of DNA research, and examples abound in other fields. The point is not that women have been cheated exclusively but that their recognition has been disproportionately omitted, even when sex discrimination has been inadvertent. The Nobel Committee's preference for discrete discoveries rather than contributions to knowledge leading up to discoveries has also limited the recognition of women, for we know more cases of men's harvesting what women have sown than the other way around. When Maria Mayer came up with her shell model, she, too, could have fallen prey to men eager to usurp her ideas. She likely would have let them, since she was self-effacing to a fault; in seminars she stopped talking no matter how rude a man's interruption. "She felt she had to be a 'lady,'" her daughter, Marianne, explained, ". . . if someone else did some work, she always let them publish it first and then she published her additions afterward. She was always very conscious that she had to behave properly if she was not to be accused of being a conniving, abrasive woman."[12]

When Mayer saw the unpublished paper of German men that suggested that they, too, independently had worked out the shell-orbit model, her impulse was to defer to them. She wrote the group leader, J. Hans Jensen: "Wigner, who does not seem to like the shell model, made the comment 'If Jensen has done it too it is probably right.'" Jensen responded: "You have convinced Fermi, and I have convinced Heisenberg. What more do we want?" Still, she delayed submitting her findings until she and Jensen could publish together in the *Physical Review* in 1950. Over the next several years they corresponded back and forth, fine-tuning their ideas for a definitive book on the shell-orbit model, but the division of work was hardly 50-50. Jensen had an "aversion against writing manuscripts" and procrastinated badly. When *Elementary Theory of Nuclear Shell Structure* was finally in print in 1955, Maria was named first author; Jensen conceded the courtesy for a work that was largely hers.[13]

Eight years later the greatest honor bestowed in all of Western science was Jensen's as well as Mayer's; they made independent discoveries and came together to reap the rewards as a team. As the first American woman to win a Nobel Prize in physics (she had become a U.S. citizen in 1933), Mayer begs the question, If she was so deferential to men in her field, why was her fate different from Meitner's, Wu's, and who knows how many other women physicists'? Why did *she* win the Nobel Prize and not the others?

A pattern emerges that may provide a clue: when we look at the women who won Nobel Prizes before 1970—the Curies, Maria Mayer, Gerty Cori, and Dorothy Crowfoot Hodgkin—we see that their close proximity to elite male scientists was a factor in their science-winning distinction.

Their experiences run counter to those of most women scientists. In general, single women, seen by employers as unburdened by a husband's relocation or children's needs, have fared better in science than married women. Harriet Brooks, an expert on radioactivity who worked with Ernest Rutherford at Cambridge, took a job, as so many women did early in the twentieth century, teaching at Barnard, a women's college, and found that even there women were dismissed upon marriage. In 1923 all the women listed in *American Men of Science* were single. Marriage and scientific career were mutually exclusive choices for women but not for men, and statistically this has remained the case. In 2009, studies confirmed that 70 percent of male science faculty were married with children versus only 44 percent of women at the same ranks. Twelve years after receiving their doctorates women were more than twice as likely to be single and significantly more were likely to be divorced. Such statistics cannot be blamed on lifestyle preferences. Forty percent of the women surveyed regretted that they hadn't had more children; in their minds, it was still not possible for them to have it all.[14]

And yet at the very highest echelons of science, where Nobels are won, marriage has seemed to be an asset. Connections to men have been crucial—not because the work of women laureates has not been worthy in its own right, but because without such connections, sexism would have prevented these women's access to the channels of visibility that their husbands provided. Margaret Rossiter confirmed that, while there were proportionately more single women in science in the 1920s and 1930s than today, a disproportionately high percentage of married women listed in *American Men of Science* were "starred" as exceptional. In such successful careers as Maria Mayer's, marriage served as a convenient social cover for professional interactions. Mayer benefited from a husband who was a scientist, but also from the social permissibility to form meaningful relationships with married male scientists who could not have worked as closely with an unmarried woman. That's not to say that her relationships with elite men were *always* enabling. Those whom Mayer knew intimately protected and propelled her, but the male establishment also found ways to exploit her. Such married scientists as Esther Lederberg, Isabella Karle, and Ruth Hubbard watched their husbands win Nobel Prizes for their joint research. Indeed, their experience suggests that marriage could in fact obscure women's contributions, unless they established a bit of distance from their mates. To understand Mayer's success is to understand how she worked out this balance, but also to qualify it and to view her failures and frustrations along the way.[15]

Part 1: Maria Goeppert Mayer: A Tale of Proximity and Patronage in Elite Science

Maria Goeppert's road to the prize started back in Göttingen, Germany, where her father was a pediatrician and university professor, and her mother a faculty wife. The Goeppert men had been university professors for six generations, and because Maria was an only child, Friedrich Goeppert expected her to continue the tradition of excellence. "There never was any question of my being able to go on to the university and study whatever interested me," she recalled, though this was taking a lot for granted at the time. Very few schools in Göttingen prepared girls for higher education, and so she moved from one private institute to another until her parents enrolled her in a suffragist-run *Frauenstudium* that trained women for college. When it closed before she had completed her studies, she took the entrance exam for the university a year earlier than boys her age.[16]

Goeppert's contemporaries Chien-Shiung Wu and Elizabeth Rona tell similar stories about their formative years as physicists-in-training, all of them attributing their early education to the unconventional thinking of their fathers. A leitmotif in the early recollections of elite women scientists is the dad who ran outdoor experiments and tinkered with gadgets with the daughter he treated like a son. He encouraged college training, even when Mom worried that her daughter's marriage prospects would suffer.[17] Sociologist Alice Rossi noted the strong father-daughter bond among world-class mathematicians, and certainly women physicists have echoed Mayer's perception that professional fathers had more to do with their scientific proclivities than had their domestic mothers. Anne Roe noted the same in *The Making of a Scientist* but went one further, blaming intervening mothers for failed scientific careers, nullified before they had begun. In this she didn't sound different from Philip Wylie or other social scientists who blamed reproducing women for postwar social ills; this time Mom was the one who prevented the potential Nobel Prize winner from saving the world.[18]

There is truth to the studies, insofar as the majority of fathers of Nobel Prize-winning scientists through the early 1970s were physicians, engineers, science teachers, or researchers in their own right. One study in 1954 confirmed male scientists' social and emotional distance from their mothers and noted that, in the rare instance of female scientists, one could find an even greater alienation from female parents. Even female science majors at Ivy League schools at the end of the twentieth century typically had fathers, not mothers, in their respective fields. But more important perhaps are women's *perceptions* that scientific talents are paternally influenced and acquired, for they reveal a culturally pervasive tendency to associate scientific ability with masculinity. Goeppert felt more affection for her father than for her socially graceful mother. "My father was more interesting," she told a biographer. "He was after all a scientist." In this she echoed her father's belief that domesticity dulled the mind and stifled the intellect. Friedrich told the parents of patients in his pediatric practice that mothers' coddling harmed children, for their fear of the larger world inhibited curiosity and risk-taking, traits essential to great leaders and thinkers. He vowed he would be a liberating influence in his daughter's life. Instead of buying baby dolls, he designed special lenses so that she might observe a solar eclipse. "Never become a woman!" he implored. A life of marriage and motherhood was not enough in his eyes. She needed interests and the ability to think as men could.[19]

Maria Goeppert internalized the message thoroughly and couldn't be bothered ingratiating herself with female classmates, none of whom went on to the doctoral degree. She socialized almost exclusively with male students, not an uncommon practice of Nobel Prize-winning women. Barbara McClintock, Rita Levi-Montalcini, and Rosalyn Yalow recalled that from early childhood they also preferred the company of males. Mayer thought that men provided stimulating conversations without the time-consuming intimacy women expected afterward. In later years she justified her preference as one simply born of convenience. "I've never had time for science, my family *and* kaffe klatsches . . . Large groups of women tend to get shrill." The more interesting challenge was engaging men: "to keep up with them was wonderful."[20]

While Goeppert didn't view her mother as a primary influence, in the end she was not a liability either, for Mrs. Goeppert hosted lavish dinner parties for Göttingen faculty that her daughter eventually matched, despite herself. Such social entertaining as well as family relationships were helpful to Maria's success. The Goeppert family was connected to some of the most revered men in Göttingen science; Max Born and James Franck were her parents' friends, as was the esteemed mathematician David Hilbert, who was mentor to Emmy Noether and who invited fifteen-year-old Maria to his Saturday lectures.

Maria Goeppert's early introduction to campus life made it less shocking to people when she entered the university formally as a student in 1924. At the time the ratio of men to women studying in any field of the German university was ten to one, a ratio far below that in the United States. In the sciences and especially in physics, women were even more rare. And yet Goeppert became friends with Arthur Compton, Wolfgang Pauli, Werner Heisenberg, Leo Szilard, Linus Pauling, and Edward Teller—all at Göttingen. In Max Born's advanced theory seminar she sat with Paul Dirac; Eugene Wigner; Enrico Fermi; John von Neumann; and Robert Oppenheimer, the insuppressible American. Cecilia Payne-Gaposchkin and Marie Curie never forgot the embarrassment of sitting in a reserved section of the physics lecture halls of the Cavendish and the Sorbonne, but Goeppert never felt the same sting of ostracism. After Born's seminar she joined the men for a walk in the hills and dined with them at rustic inns. Born swore that she got through his courses "with great industry and conscientiousness, yet remained a gay and witty member of 'Göttingen society.'"[21]

The responses that Maria Goeppert elicited from male scientists were very different from those provoked by prominent female contemporaries. Barbara McClintock exuded a pixielike androgyny in the genetics lab at Cornell, and Annie Cannon presented herself as the hospitable matron of the Harvard Observatory. Lise Meitner, too, was devoid of sexual allure and remained single all her life—"married" to her research, as male colleagues saw it. But men in Göttingen thought Goeppert radiated youth and femininity, with her slim figure, strawberry hair, and lively blue eyes. She laughed, she danced, and she fraternized. Victor Weisskopf and Robert Mulliken admitted to being seduced by her charms. "For most of these scientists, they'd never met a woman who was as intelligent as they were," her daughter explained, adding to her myth, "and then to have a woman who was very definitely a woman—it was an unbeatable combination." Some wondered why Born took so much interest in the young, attractive pupil, but their relationship was like so many she cultivated throughout her life. Her doctorate examination committee consisted of Born and future Nobel laureates James Franck and Adolf Windaus, men whose reputations intimidated those not already made timid by the intense *Doktorvater-Studentin* apprenticeship of the German university. Graduate students deferred to their masters to the point of worship, yet Goeppert won their affection and respect.[22]

In 1927 Goeppert met an American, a Rockefeller Foundation fellow named Joseph Mayer, who had come to study quantum mechanics under Franck. The son of a bridge engineer, he had gone to Caltech and Berkeley and had cultivated an easy manner that defied the stereotype of the cloistered lab rat. To his German peers he was a man of the Jazz Age: his dress was stylish and he indulged in the local nightlife. He came to the Goeppert home in 1927 to inquire about a room: "I went and rang the doorbell, and a pretty little snip of a girl came to the door, and wouldn't talk German to me. Her English was perfect." She was "a terrible flirt," he recalled, "—but lovely, and brighter than any girl that I had ever met." Before long they were out dancing almost every night.[23]

Like Goeppert's father, her future husband had little tolerance for uninspired women; he loved that she needed the stimulation of science. When he asked her to marry him in 1929 Joe Mayer didn't go on about how happy she would make him. He would make *her* happy, he promised, since he wouldn't make her give up physics. Maria admitted years later that she likely would not have continued toward the doctorate degree without Joe's prodding her. On Christmas Eve, as she worked in the kitchen to prepare a holiday meal for friends, she realized that she desperately missed the maid. "If you go on in science," he told her, "I will always keep a maid for you." When she procrastinated about writing her dissertation, Joe took her to the Netherlands to talk the thesis through with the great Paul Ehrenfast, whose solution was to lock her in his guest room and force her to write. Goeppert sat in the room alone, admiring the walls, which had been signed by famous scientists who had stayed there. Einstein's signature was prominent as one of Ehrenfast's closest friends. Having grown up in Göttingen, Goeppert had taken for granted the privilege of watching the quantum revolution emerge before her eyes. Her proximity to luminaries was unique, and she had the natural gifts to move among them. "I solved my problem within an hour," she remembered. "It was the basis for my doctoral thesis." Eugene Wigner called the work "a masterpiece of clarity and concreteness." Decades later, with the greater light intensity of lasers and the development of nonlinear optics, physicists confirmed much of what she had hypothesized in 1929.[24]

Maria married Joe in January 1930 and got her PhD two months later. She was twenty-four years old, and, like other women, couldn't get a job in the academy that had trained her. Margaret Maltby, the first woman to receive a PhD at Göttingen, or any German university, ended up leaving Germany to head up the physics department at Barnard, where she wasn't promoted to the rank of associate professor until the age of fifty-two. Emmy Noether took an unpaid appointment at Göttingen at the age of forty-one but then was fired and forced to leave for the United States, where she secured a job at Bryn Mawr, a women's college. Lise Meitner and Hertha Spöner, Göttingen's second privatdozent after Noether, also went to the United States to secure academic positions. Mayer followed precedent: she and her husband emigrated to Baltimore, where he assumed a tenure-track position at Johns Hopkins.[25]

No one there had her expertise in quantum mechanics, but the university was no different from nearly all coeducational colleges throughout the United States, particularly during the Depression, in its refusal to hire both husband and wife into full faculty positions. Antinepotism policies were ostensibly implemented to prevent the hiring of unqualified spouses into positions better occupied by other applicants, yet they invariably quashed the careers of qualified women married to similarly or less-credentialed men, rather than the other way around. Faculty wives typically fell into unpaid assistantships or "volunteer" positions that gave them access to facilities in exchange for heavy teaching loads. They worked in labs that resembled male fiefdoms and felt fortunate to get credit on publications or assurances of continued employment. Mayer was grateful to get workspace in the attic of the physics building, where she took on graduate students, Robert Sachs being her devoted first. Unofficially, she was paid a few hundred dollars a year to help faculty with their German correspondence, and over time she gave graduate lectures on a "voluntary" basis. Her courses were listed in the university catalog under "G," a single initial for her maiden name.[26]

At Hopkins, Mayer had the professional status of a candy striper, but she chose to look on the bright side. It was a rare opportunity to be a fly on the wall, taking in all she could about as many areas of physics as possible. Again, she was in the presence of stimulating men. James Franck had come to Hopkins, and Edward Teller and George Gamow were in Washington, convening seminars she attended. She teamed up with experimentalists and men of the math department. Karl Herzfeld, a theoretician of kinetics and thermodynamics, helped Joe Mayer move her into physical chemistry. She collaborated with Herzfeld's graduate student Alfred Sklar, applying methods of group theory and matrix mechanics to the structure of organic compounds. Soon she was publishing with Robert Sachs on meson exchange.[27]

Her dealings with students reveal her unique status as a woman physicist. All were young men and in awe of her command of theoretical concepts, but they also found her more approachable than they did the great men in their field. Sklar admitted that he felt more comfortable asking her for letters of recommendation or tips on real estate than he did the male faculty; she wasn't the father figure evaluating his every move. He wrote her in 1944: "'Mrs.' doesn't fit; 'Dr. Mayer' is not you; Dr. Goeppert Mayer is too formal for words. Most people I know solve the dilemma by resorting to Maria." He couldn't imagine, however, calling her husband anything but "Dr. Mayer."[28]

The volunteer physicist had proved her indispensability in the department and had found advocates among her colleagues. Herzfeld thought her so ruthlessly exploited by the university that he paid her out of pocket a modest compensation. When he requested her name be added to the departmental letterhead, the dean of physical sciences took offense and removed all names but his own. Mayer paid no mind to the politics; without a formal appointment or voting privileges she felt free to leave campus when the need arose. She spent her first three summers away from Hopkins in Göttingen to resume work with Max Born, a collaboration that resulted in a groundbreaking paper in the *Handbuch der Physik* in 1933. She developed theories of double beta-decay based on Fermi's formulation of nucleon-neutrino interaction and published on the excited electronic states of the benzene molecule. Colleagues were struck by her intellectual range.[29]

Life in Baltimore was not all science; she hosted a steady stream of anti-Nazi defectors in her home and had a baby in 1933. Mayer morphed into the doting first-time mother her father had vilified. She took time off from science during her daughter's infancy to be the full-time parent her mother had been for her, content to keep up with the literature at home. But during her second pregnancy with Peter in 1937, she spent her final trimester at home and was struck with the same epiphany she had had making Christmas dinner in 1929: that she really had little tolerance for domesticity. She learned to loathe coming up with daily menus and humoring neighborhood mothers with empty chitchat. Her frustration mounted whenever she had to leave campus to nurse sick babies at home. When Marianne was a newborn, Mayer had gradually worked back up to a one-third-time teaching schedule. Immediately after Peter was born, however, she went back to work "with a vengeance."[30]

Joe Mayer made sure during the second pregnancy that she had the stimulation of a project to occupy her—a textbook on statistical mechanics that he worked on with her at night. Benefiting from the editorial eyes of both a chemist and physicist, *Statistical Mechanics* became an instant classic, the most comprehensive work in the field then and for decades afterward. It also had the unintended effect of catapulting Joe's career and doing little for Maria's, for those who didn't know better assumed that she had been an editorial assistant rather than a coauthor. Joe accepted an associate professorship at Columbia in 1939 at twice his Hopkins salary, hoping the change would also benefit his wife. Harold Urey, the chair of Joe's new department, rustled up the title of "lecturer in chemistry" for Maria, since the publisher of *Statistical Mechanics* couldn't stomach printing a book cover with no academic title following her name. The unpaid position was all she could get at Columbia for the first two years. She applied for a full-time job in the physics department, but when the chair refused her, she settled for office space and unpaid teaching assignments.[31]

The move to New York seemed to be unrewarding for her, but then Max Born assured her from overseas that her underemployment wouldn't lead to boredom, since Urey, Teller, and Fermi were there. Mayer enjoyed these men on campus, but they were also her neighbors in Leonia, New Jersey, the suburb where she and Joe settled with the kids. On any given weekend Teller shot spitballs with Peter Mayer, while Fermi talked history with Marianne; everyone's kids played together. With the help of caterers, Mayer extended hospitality to colleagues who convened in New York for science meetings. Graduate students who couldn't afford hotels found it most invigorating to talk physics at the Mayers' breakfast table with luminaries in the field. In offices on campus or on sofas in suburban dens, Maria Mayer chain-smoked with men and ruminated on the mysteries of atoms.[32]

Some of their American-born spouses were unimpressed. Laura Fermi noted the tensions when Maria Mayer talked "technical" with the men, and couldn't understand why Americans separated the sexes in social settings. Husbands went to stag parties and left "poor young wives to mope at home" and then expected their women to plan luncheons with faculty wives who were virtual strangers. Maria Mayer avoided the Tupperware parties and became the focus of envy and disgust. Joe wondered which women had intervened to have her excluded from the dinner that followed his department's weekly seminar. Maria urged Joe to say nothing about it and happily accompanied Edward Teller to the opera instead.[33]

In 1941 Mayer was honored with election into the American Physical Society, but her status at Columbia didn't change. She accepted a part-time salary at Sarah Lawrence College to teach math and physics courses to nonscience majors and soon taught double the load of introductory courses and chaired the entire science department. For all the responsibility, she was unable to secure a permanent contract or full pay. Max Born remained encouraging: "Your influence will lift the standard of science amongst the wealthy girls of Sarah Lawrence skyhigh." By all accounts it did. Administrators worried that her highbrow theory wouldn't prepare students for their more likely need to regulate the flue on a home furnace. Mayer asked them if women learned English only to read cookbooks. It was the beginning of the atomic age, and she had every intention of teaching the same lessons in nuclear physics that she taught to men at Columbia.[34]

Mayer's teaching was interrupted when the United States declared war on the Axis powers and physicists were called to defense work, regardless of their sex. As a Manhattan Project manager, Urey carved out for Mayer a position in Columbia's Substitute Alloy Materials (SAM) Laboratory. The appointment had the potential to thrust her into a position of visibility, as war work did for young male physicists. She was better compensated than ever before and had a team of researchers who answered to her. But the war's timing coincided with teaching and the rearing of small children. She arranged for the English nursemaid to work longer days but insisted on being home to read to the kids before bedtime and on being with them on the weekends. Urey readily agreed, assigning her the more peripheral work of isotope separation by photochemical reactions. Soon Edward Teller asked Mayer to join his Opacity Project, and she agreed to take it on. Looking back, her daughter thought it a good thing, since her dad was at Aberdeen Proving Ground doing weapons research, and her mom was "actively miserable without him," her only solace to be completely immersed in work.[35]

After the war Mayer returned to teaching at Sarah Lawrence and spent time tutoring Marianne in German and addressing Peter's poor performance at school. Her days were full, but she wanted more meaningful science. Joe had just received word that he had been elected to the National Academy of Science and offered a position at the University of Chicago; again, he hoped for his wife that the change would be for the better. Everyone who was anyone in atomic physics was moving or had moved to Chicago, including most of their Leonia neighbors. But nepotism policies proved equally stringent there. Mayer was given faculty status, but again no pay, even though she worked in a separate department from Joe's. She was a "volunteer associate," which meant that she handled committee work, teaching, and advising duties in the physics department for free. Nevertheless, Robert Sachs, now the head of the physics division at Argonne, offered her simultaneously a part-time position as a senior physicist. Mayer was pleased with the spacious office and "nice consulting salary," but more important, she no longer felt like an appendage. She enjoyed unprecedented access to resources and took solace in knowing that faculty friends were fighting for her to earn a full-time salary.[36]

Mayer's life in Chicago was nothing like Marie Curie's anti-natural path. She threw the grandest of parties in their three-story South Side mansion, decorating lavishly and entertaining hundreds of people. Younger scholars looked up to the Mayers as the epitome of sophistication. For Maria the hospitality was effortless, and she was also the easiest of guests. Before a symposium on high polymers, colleagues convened at a Kyoto inn and drank sake as geisha girls danced around them. Women's designated work at the event was as entertainers, for female chaperones escorted the scientists to the public baths. And yet Mayer was, as always, too graceful to appear out of place. When colleagues threw stag parties, she wore slacks and joined them. Male scientists consequently decided she was one of them, electing her to the Akademie der Wissenschafter in Heidelberg and the National Academy of Science. She was the fifth woman to receive the latter honor—a decade after Joe had been elected. She collaborated with Jacob Bigeleisen on isotopic exchange reactions and plugged calculations into ENIAC, the first electronic computer, to figure out criticality for a liquid metal breeder reactor. Some of her fondest memories were of the seminar at the Institute for Nuclear Studies, where Teller sparred with Fermi, Urey talked of the moon, and Willard Libby announced his discovery of carbon dating. She also fortified personal and professional bonds with the quantum theorist Gregor Wentzel, soon to be the father-in-law of her daughter, Marianne.[37]

This was the milieu in which Mayer conceived her shell orbit theory. Men liked her; some loved her, and most were protective of her. Joe Mayer, Fermi, and Teller were three who insisted that she publish alone and immediately, so that her ownership of the theory would never be questioned. Their instinct on her behalf was to strike while the iron was hot. Joe grew so exasperated that he eventually lost his temper: "For God's sake, Maria, write it up!" But she remained reticent; Fermi should be a coauthor, she argued, since he had asked the question that unblocked her thoughts. Fermi refused to see the logic. Were his name anywhere on the paper, he would get all the credit, he promised her, and he had no intention of undermining her efforts. Her model explained irregularities of the stability in nuclei, as well as all the phenomena of nuclear spectroscopy, including beta and gamma decay. Before his death in 1954, Fermi took it upon himself to nominate Mayer for the Nobel Prize.[38]

Mayer's years in Chicago were happy ones, but they ended in 1960 when she was lured away. The deciding factor was not her lack of formal status, but rather that the men she knew had left—Teller in 1952, followed by Libby in 1954, and Urey in 1958, when he accepted a position at the University of California at San Diego (UCSD). Administrators throughout the California system wanted big names to build up its nascent science programs and were amenable to hiring married couples if necessary. Urey convinced them that his longtime friend would bring stature to UCSD, and in 1959 Maria received an offer of full professorships for her and Joe both. She didn't object to her half pay for nine months to Joe's full twelve-month salary; the money was far better than what she had received anywhere else, and the official appointment would allow her to earn more through grants and contracts. Because the position was tenured, she would not have to think about contract renewals. Urey occupied an office near hers in the chemistry-physics building. The California climate was ideal for resuming her cultivation of orchids and joining Joe for lunchtime swims at the beach.[39]

At four in the morning on an early November day in 1963 Mayer got a call from a Swedish journalist, who was the first to tell her that she had won the Nobel Prize for her shell-orbit theory. Joe poured champagne and the two sat on their deck eating bacon and eggs until the sun came up, enjoying the quiet before the storm. The phone rang incessantly for the rest of the day as the media converged on them. Bumming cigarettes from the journalists, Maria was able to get through many hours of lights, cameras, and interviews with composure intact. Urey managed to sneak past reporters to give his friend a tight embrace. "Life will never be the same," he said in her ear. Joe left to sit on a chemistry exam committee and returned wearing an understated short-sleeve shirt and cotton trousers. He had no intention of upstaging his wife at her finest hour. As the press moved to the lawn outside, the two Mayers stood arm in arm answering questions. A journalist asked Joe whether he thought of Maria as a wife or a scientist, as if they were mutually exclusive categories. Incredibly, Joe bothered to answer: "Why a wife, of course, and a very wonderful one." More fiercely than at any other time, he felt compelled to defend her normality as a woman.[40]

Mayer received letters and telegrams of congratulation from scientists all over the world, as well as from local schoolchildren, who were excited to be near the "San Diego mom" who had won the prize. In one letter, a local woman invited Mayer to stop by some day after three o'clock, since her fifteen-year-old daughter would be impressed to meet a female Nobel Prize winner: "I'll make us some coffee + we can smoke together." The letter seems inappropriately familiar, but then again Mayer's letters from male graduate students also suggest that they had always felt more comfortable approaching her than they did men of her professional stature. As a woman, she was a scientist stripped of all that made scientists imposing. Sometimes she felt like an impostor, as if her Nobel Prize had been won fraudulently—or at least fortuitously. When a fifth-grade girl wrote to ask how a woman could have achieved so much in science, Mayer replied: "I have been very lucky in my career. Lucky in coming to America when I did, lucky in meeting my husband, and lucky in my choice of research." This was not a whitewashed answer for a child's ears. Lucky was truly how she felt.[41]

Maria Mayer and Gerty Cori:
Rethinking the Nobel Success Story

It took a Nobel Prize for Americans to understand what a handful of elite scientists already knew about Maria Mayer: that she was one of the most brilliant theoretical physicists in the world. In 1896 Alfred Nobel had left his fortune to fund prizes in the burgeoning fields of physics, chemistry, and medicine, as well as literature and peace. That he united those who made technology with those who made peace testified to his belief that science could and would improve the human condition and that its greatest practitioners deserved any messiah complex that might result from their accomplishments. When psychiatrist Helen Tartakoff coined the term "Nobel Prize complex" to describe certain narcissistic and obsessively ambitious patients, she undoubtedly had scientists rather than poets or peace activists in mind. Each December dignitaries convene in the Stockholm Concert Hall, dressed in formal evening attire. After the Royal Hymn the new class of laureates files to the left of the stage to accept gold medals, on the backs of which are the female forms of Natura and Scientia. The physicists walk across the stage first, followed by the chemists, the medical researchers, and the nonscience laureates. In 1963 Maria Mayer was a rare woman heading this ritualized processional.[42]

For each Nobel laureate there have been thirteen members elected to the American Academy of Science and twenty-six listed in *American Men and Women of Science*. Since 1901 the Nobel Prize has been the mark of supreme honor in Western science, and no more so than after World War II. Although economists and writers can also win the prize they understand that their selection may be more subjective than the selection of scientists. Perhaps the difference in perception stems from the positivist tradition of science itself. Scientists are trained to think of phenomena as measurable, of preeminence as empirically knowable. And thus it follows that, as a culturally accepted measure of greatness, the Nobel in science generates authority for its own selection process and for those bestowed with the prize. Upon winning, scientists discover that their opinions hold greater weight, whether in reference to lab work or world peace. After Maria Mayer won her Nobel, she was asked to show support for causes ranging from population control to Lyndon B. Johnson's presidential campaign.[43]

It is not news that Nobels are rare, or that they have been rarer for women. Out of the hundreds of prizes awarded to scientists, we can almost count women Nobelists on two hands, thanks only to the selection of more women of late: Marie Curie (1903, 1911), her daughter Irene Curie-Joliot (1935), Gerty Cori (1947), Maria Mayer (1962), Dorothy Crowfoot Hodgkin (1964), Rosalyn Yalow (1977), Barbara McClintock (1983), Rita Levi-Montalcini (1986), Gertrude Elion (1988), Christiane Nüsslein-Volhard (1995), Linda Buck (2004), Françoise Barré-Sinoussi (2008), Elizabeth Blackburn (2009), Carol Greider (2009), and Ada Yonath (2009). Of these women, even fewer have won prizes without sharing them with husbands, male collaborators, or men whose unrelated accomplishments have merited distinction at the same time. In the rare instance when a woman has won a Nobel, many assume that her dedication allowed her to perform outside her female skin, or, if she is not truly exceptional, that she positioned herself strategically on the coattails of worthy male recipients. Marie Curie suffered the stigma of being described at both extremes.

In either case the underlying assumption is that normal women don't possess the goods to win Nobel Prizes. And thus, in the few instances in which women have won them, their scientific accomplishments have been less newsworthy than their eccentricities, contradictions, and transgressions as typical women. The headlines tell the story: "La Jolla Mother Wins Nobel Prize"; "She Cooks, She Cleans, She Wins the Nobel Prize"; "The Chemistry-Minded Mother"; "A Nobel Woman's Hectic Pace"; "British Winner is a Grandmother"; "At Long Last—a Nobel for a Loner"; "Winner Woman"; "An American Mother and the Nobel Prize—a Cinderella Story in Science."[44] Although most people on the street know Marie Curie won the Nobel Prize, it's unlikely they can tell you what she won it for. Nobel Prize-winning women are the news, not the science they command.

Nobel laureates are popularly recognized as part of a "knighthood" or a "science aristocracy" for whom, as for any elite class, access to resources has been integral to maintaining status and passing it on to professional progeny. Sociologists confirm that women's historical inability to "attend superior training centers," work as "apprentice[s] for master scientists," and occupy "facilities to carry out their research ideas" has led to their virtual omission from this rarified caste. In 1979 Jonathan Cole underplayed the importance of sexism in women's poor standing in elite science, but he couldn't deny that inertia was working against them: "If women are more apt to be in such disadvantaged positions than their male counterparts, then the career histories of men and women of science may be explained in part by processes of accumulation of advantage and disadvantage. . . . the 'rich getting richer,' but also 'the poor getting poorer.'" Men inherit more than their mentors' connections: they gain also their mannerisms, values, and swagger as members of the professional elite. Once they have achieved modest acclaim, their pedigrees prevent derailment off successful paths. Robert Merton termed this self-perpetuated success the "Matthew effect." Along with pedigree, the greatest predictors of Nobel Prizes have been the multiple honors—the Laskers, MacArthurs, inductions, and honorary degrees— a man and his mentors have already received. Recognizing the all-too-frequent correspondence that this dynamic has to gender, Margaret Rossiter has exchanged Merton's biblical Matthew for "Matilda," as in the suffragist Matilda Joslyn Gage, who believed that Christianity, like science, perpetuated ideas that allowed men to reap what women had sown. As an obscured figure in her own right, Gage herself was a victim of the "Matilda effect."[45]

Indeed wealth begets wealth in the rarified world of science, and inheritance historically has been patrilineal. Lists of Nobel winners reveal generational patterns: fatherly lab directors followed eventually by their male graduate and postdoctoral offspring. These have been genealogies in both the figurative and literal sense, since several younger laureates have been related to former winners. Aage Bohr won the physics Nobel fifty-three years after his father; microbiologist Frederick Robbins won eight years after his father-in-law won in chemistry. Alan Hodgkin, a Nobelist in medicine, married the daughter of American laureate Peyton Rous; and Paul Dirac, a laureate in physics, married the sister of another physics laureate, Eugene Wigner, who married the sister of his colleague John Wheeler. The list of inbreeding goes on, suggesting that women have been integral to forging the social bonds that create laureates even when they have not been laureates themselves.[46]

Women in the immediate postwar years could only envy the social advantages of and proximity to mentors that male peers enjoyed in and outside the lab. Thanks to the circumstances of birth and marriage, Maria Mayer was a rare woman who had access to such people. Biochemist Gerty Cori, a winner of the Nobel Prize in 1947, suffered institutional sexism but was also buoyed by a supportive husband who considered her his intellectual equal. Carl and Gerty Cori pioneered the study of the hormonal regulation of blood sugar and the role of insulin in the body's regulatory system, and by all accounts they also enjoyed a most compatible marriage. Carl simply refused to work without his wife, and thus, when sexist and anti-Semitic sentiments prevented her from getting research jobs in Europe, he charted a professional path for them both in the United States, first at New York's State Institute for the Study of Malignant Diseases, and then at Washington University in St. Louis. His decision to work with his wife frustrated his superiors and delayed his own promotion. He refused a tenured position at the University of Rochester, for example, when he discovered that Gerty couldn't work with him in the lab. The department chair chastised her for getting in her husband's way. It was simply "un-American" for a woman to encroach on her husband's turf, he told her, but Carl hardly thought his wife a liability. Her hand in elucidating the process of glycogen metabolism opened doors that he took advantage of on her behalf.[47]

Antinepotism policies and sexist bias conspired to keep Gerty Cori in low-paying positions with little prospect of advancement, but she never complained. She wasn't envious of Carl, especially not when his promotions led to more administrative tasks outside the lab. This was a woman who rushed back from the birth of her son so that she could continue isolating glucose-1-phosphate, soon to be known as "Cori ester," in order to trace it to the breakdown of polysaccharides. With infant and housekeeper at home, Cori continued to work full-time and finally was promoted to research associate in 1937, to associate professor in 1943, and to full professor of biochemistry in 1947, sixteen years after Carl had become a full professor and months before receiving her own Nobel Prize. Her promotional ladder was decidedly longer than for men with identical credentials and talent. Awards poured in just as she was diagnosed with a chronic bone marrow disease that was, like Marie Curie's malady, probably triggered by exposure to radiation. When her hemoglobin fell to lethal levels, Carl stopped smoking to make it easier for her to quit, too. When she became too weak to walk the corridors, she called out for "Carly," who picked up her frail frame and carried her down to lunch. The two were inseparable until she died in 1957.[48]

In tales of Western science, the Nobel is the fairytale ending. Few historians have written about Cori's professional struggles before the prize or the pain of chronic illness that quickly followed. Feminists, too, have chosen to highlight the Nobel Prize as the pinnacle of the lives of Marie Curie and Maria Mayer without qualified discussion of how events that preceded and followed gave meaning to their lives. The memoirs of male laureates are formulaic: one's professional coming of age reads as a male rite of passage, entailing ascent on a promotional ladder through prestigious labs in one's twenties and thirties. Wives accept their husbands' lengthy work hours and their meager and sporadic participation in family life. Maria Mayer and Gerty Cori's twenties and thirties were stimulating years in which they, too, felt enraptured by science, but Mayer also felt inadequate as a mother, and Cori held motherhood off until she was forty to avoid similar feelings of guilt. Unlike men, with their linear progress towards scientific prestige, women scientists may feel torn by competing interests and plagued by institutional constraints.

When one considers women scientists, then, one must think afresh about where greatness begins and how it ends. Maria Mayer and Gerty Cori were like many women who stagnated on promotional ladders or who knowingly derailed careers to tend to children. Sometimes accolades for these women would come later than for their male colleagues. Perhaps they would never receive salaries comparable to their husbands', or promotions, grants, and honorary appointments until prizes proved their eminence, not the other way around. Mayer was the lone woman in a physics dynasty that produced forty-three male Nobelists; Cori, the only female among thirty-nine other men.[49] No one disputes their intellectual abilities, but given women's grim prospects in post-war science, it's likely that blood and marriage worked together in their favor. Neither woman would have come to the United States, let alone secured work as a scientist of any kind, had she not had male support. Most women could not counter the sexism of scientific institutions. By comparison, the careers of Mayer and Cori were charmed.

Part 2: Single and Snubbed: Rosalind Franklin and the Politics of the Nobel Prize

In June 2003 James Watson, the president of Cold Spring Harbor Laboratory on Long Island, was a featured guest on the Public Broadcasting Service-syndicated *Charlie Rose Show*. Rose and Watson had lots to talk about. A discovery that would serve as the pinnacle for most careers was just the prelude to Watson's. At twenty-five he had figured out the double helix structure of DNA, and in his thirties won a Nobel Prize; by his forties he was a best-selling popular author and since then had earned honors and appointments and had become an integral part of the Genome Project. In 1990 *Life* magazine listed him and Jonas Salk as the only biologists on its roster of the one hundred most important Americans of the twentieth century. Watson had just filmed a retrospective of his fifty years of research in a documentary film, with a companion book on DNA. Rose could have begun anywhere, and yet he chose to revisit some nitpicky details of Watson's Nobel-winning discovery in 1953:

CR: If you hadn't done it, who would have done it?

JW: Probably Rosalind Franklin or Maurice Wilkins.

CR: . . . Did Rosalind deserve to share the Nobel Prize?

JW: Well, you could argue it either way.

CR: Which way do *you* argue?

JW: (Pause) . . . Well . . . you . . . it was impossible—you could only give a Nobel Prize to three people and there were four of us, so . . .

CR: Who made the decision, the Swedes [Nobel committee]?

JW: Yes. Well, Rosalind was dead, so arguing over a hypothetical thing was . . . was she a good scientist? I'd like to say yes.

CR: Absolutely.

JW: Did she find the structure? No.

CR: Did her photograph—whatever you call it—contribute to it?

JW: Oh, sure, because when I saw it . . . It had to be a helix. . . . Rosalind at that time didn't want to think of helices because Wilkins said it was a helix and you know there was conflict between the two of them.

CR: But he's the one that showed you the photograph, right?

JW: Yes . . . but it wasn't taken the day before, it was taken at least six months before. . . . It was a beautiful picture; it need not [have] been that good for me to have . . .

CR: If you had not found it . . .

JW: She should have found the structure by the time I arrived in England.

CR: She *should* have?

JW: Yeah, because she should have built models. You know, it might not work, but try it. She wanted her approach to work. We'd try any approach. . . . I just wanted the answer. . . . We didn't think of it in those days as Rosalind's problem. I thought it was Wilkins's problem . . .

CR: Everyone wants you to speak about Rosalind, it seems to me.

JW: Yeah.

CR: They almost want you to say, "Well we overlooked, we didn't give her enough credit," seems to be what people want you to say.

JW: We didn't even know Rosalind. Rosalind didn't want to see me. She didn't like me.

CR: Why didn't she like you?

JW: Probably in part because she saw me believing in helices, that's why . . .

CR: Was she discriminated against because she was a woman?

JW: No, I don't think so.[50]

Watson seemed happy to change the subject, but not surprised by the questions; journalists have asked him to defend his discovery of the double helix and to explain Rosalind Franklin's role in it many times. The prevailing narrative in the minds of Americans who know of Franklin is a victimology, a story of a woman cheated of her just rewards. She was an English scientist, not an American, but her story has been converted into a transatlantic tale of sexism that has captured the imagination of

American feminists more completely than that of any scientist in the United States. In the twenty-first century Brenda Maddox has called her "the Sylvia Plath of molecular biology, a martyr whose gifts were sacrificed to the greater glory of the male."[51] If Maria Mayer's tale of proximity to men is a success story, Franklin's is a cautionary one about what happens when proximity is lacking. The events before and after the discovery of the double helix are worth revisiting for the light they shed on the gendered politics of elite science in the postwar years. How significantly can we attribute Franklin's fate to the masculine traditions of institutional science? Had she been married to a male scientist or protected by male mentors, would events have transpired differently?

Feminist biographers remind us that Rosalind Franklin had the natural talent and masculine ambition to win the Nobel Prize. She was born to an Anglo-Jewish banking family in London in 1920, and like her brothers, she quickly learned to tinker with gadgets. By the age of six she excelled in mathematics. Rather than discourage her from career paths, her parents enrolled her in one of the few English schools that took the occupational training of girls seriously. By sixteen, she had chosen to pursue science. She enrolled at Newnham, the women's college of Cambridge, and graduated with a degree in physical chemistry in 1941. The Cavendish Laboratory of Cambridge was home to some of the most esteemed scientists in the world, but Franklin did not find that her associations with elite men translated into ready opportunities in the work world, largely because no one took the initiative to take up her cause. She may not have gone further in science except that the British war effort needed scientists who understood the properties of coal in order to maximize its use. She secured a post at the British Coal Utilisation Research Association as an assistant research officer. Her findings there became the foundation of five papers and a PhD thesis that she submitted at Cambridge in 1945. At the age of twenty-six, she had become one of the premier coal chemists in the world.[52]

One of Franklin's early influences had been Adrienne Weill, a physicist who had worked with Marie Curie and who had left France to avoid the Nazis. A forebearer of modern feminists, she had a daughter but no husband, a sophisticated style, leftist tendencies, a penchant for existential philosophy, and an unshakable sense of who she was and wanted to be. She helped to cultivate Franklin's love of French people and culture and secured a job for her in the Paris lab of Jacques Mering, who helped Franklin hone the X-ray diffraction techniques that defined her career thereafter. Franklin thrived in Paris. There Simone de Beauvoir could ruminate on philosophical questions without suffering judgment of her unconventional sex life, and there, too, Franklin could be a respected professional with no love life at all.[53] She developed expertise in crystallography, a lab-centered field that was exceptional for its historic embracing of women. Sir Lawrence Bragg, the originator of the field, welcomed women into his Cambridge labs during decades when they were to be found nowhere else in the university. While the percentages of women in crystallography historically have been higher than in other physical science fields, they haven't been high compared with those in biological or nonscience fields, and thus the characterization of crystallography as "woman's work" likely has had much to do with men's perception of it as a "service" to other science disciplines—as phlebotomy is to medical research or as "computing" was to turn-of-the-century astronomy.[54]

Crystallographers seemed to be the helpmeets of other scientists, and in 1950 good ones were indispensable. J. T. Randall offered Franklin a three-year fellowship to analyze protein structures in his biophysics unit at King's College, London, though privately, lab chief Maurice Wilkins was urging Randall to make the structure of DNA Franklin's primary area of concern. Chemists had already identified the protein bases that made up DNA and had figured out the amino acid sequences for insulin polypeptides. Caltech chemist Linus Pauling was months away from announcing his discovery of the α-helix. Wilkins was abreast of these developments and felt that he was on the precipice of a major discovery in a field that was innovating daily. He needed someone qualified to capture the X-ray diffraction images of DNA so as to identify the mechanism transferring genetic information. Randall was confident that Franklin could provide what was needed.

But immediately there were misunderstandings. By virtue of Randall's assigning Franklin to DNA, Wilkins assumed that he was in charge, but Franklin had been told otherwise. Raymond Gosling, a graduate student assigned to X-ray work before she had arrived, recalled their introductory meeting. Randall first handed him over to Franklin, then showed her the X-ray diffraction pictures taken to date and told her to resolve the structure. Thus Franklin thought she was heading up a small and independent team of two. Tensions rose immediately.

The more Wilkins tried to tell her what to do, the more Franklin dug in her heels. Wilkins found her "fierce" and overpowering in arguments. Lines of communication deteriorated and then completely broke down. Wilkins complained of her unwillingness to collaborate; she was a lone wolf who hoarded her data, a woman without the grace to concede points when she started to look combative.[55]

If Wilkins found Franklin blustery, it was partly because she had few outlets for airing frustrations. King's, like Cambridge and like English science generally, was a gentleman's club to be navigated, not the community she had experienced in Paris. When she joined the biophysics unit, less than 1 percent of the British Royal Society was female; the marginalization of women reverberated down into their day-to-day activities in lab settings. Of the thirty-one scientists on staff in the biophysics unit in 1952, eight were women, and only one or two worked with Franklin directly. Forbidden entrance to the common rooms where male staff smoked and took their coffee, Franklin often ate in the student hall or left campus altogether. Randall was kind, but others were misogynists. "She was used to a civilised intellectual life, discussing painting, poetry, theatre and existentialism," a graduate student explained to Brenda Maddox. "Now she found herself among people who had never heard of Sartre, whose chief reading was [the] *Evening Standard*, and who enjoyed 'the type of girls that would get drunk at departmental parties and be passed from lap to lap having their bra undone.'" Franklin put her head down and tended to her experiments. Men mistook her discomfort for haughtiness, severity, an appalling lack of femininity. She refused to flirt or suffer the pranks pulled at her expense—so she was declared an outsider from the start. Men patronized her to make it clear that they doubted her scientific abilities.[56]

Enter the young and self-propelling James Watson. A prodigy of sorts, he had been featured on the national radio program *Quiz Kids*, had entered the University of Chicago at fifteen, and had earned a PhD from Indiana at twenty-two. The literary heroes of science enthralled him; his dreams of emulating Martin Arrowsmith and Paul de Kruif's *Microbe Hunters* kept him motivated, and Erwin Schrodinger's *What Is Life* turned his focus to the workings of the gene. In Bloomington he studied the effects of X-rays on bacterial viruses with Salvador Luria and Max Delbrück, men who, ironically, won Nobel Prizes after he had won his. He worshipped Delbrück and tried, like a good disciple, to position himself next to him at meals in the hopes of taking away nuggets of wisdom. A friendship developed over tennis and beers after summer seminars at Cold Spring Harbor. Luria, an Italian, brushed up on his English by rewriting Watson's doctoral thesis. Looking back, Watson was sure that one of the keys to his success had been removing neckties and using first names early—being seen as an equal to important men from the start. Men formed bonds over cognacs or while grilling in the backyard, but women were shunned when they seemed to mix domesticity and lab work. For young Franklin, informal relationships had not been seen as improper in Paris, but they were viewed so in London. Her relationships with possible advocates at King's remained stilted and strained.[57]

Franklin had no mentors, while Watson's exuded influence enough to cross the Atlantic. Luria cleared paths to his success, securing for him a National Research Council fellowship and the opportunity to get crystallography experience with John Kendrew and Max Perutz at the Cavendish. Perutz's graduate student was the garrulous, thirty-five-year-old Francis Crick, the resident theorist of the unit. He and Watson bonded immediately, perhaps because each fueled the other's designs on DNA structure. Crick explained X-ray diffraction to Watson and introduced him to Maurice Wilkins in October 1951. Wilkins explained to Watson that the X-ray patterns captured at King's suggested that DNA would consist not of single polynucleotide chains but, rather, of two or three helical chains intertwined and bonded to each other. Franklin was the one with the latest data, Wilkins told Watson, and she was presenting her findings in a month's time. Watson made sure to attend her presentation, convinced that unlocking the structure of DNA was his ticket to the Nobel Prize.[58]

Lawrence Bragg, director of the Cavendish, made it clear that DNA was King's problem, but Watson and Crick thought the discovery of the structure a masculine race to be won. To lessen the look of impropriety, Bragg later described the discovery as a combination of Wilkins's patient plodding and Crick and Watson's "rapid fire solution," passing over and ultimately ignoring that Franklin was the only scientist in the world who had consistently been obtaining data on the structure during the two years that led to its discovery. Crick and Watson built models of DNA in a trial-and-error fashion, bouncing ideas off each other until some of them stuck. Their process was social; they talked about

DNA over tennis, at the pub, and during lavish dinners Odile Crick prepared for them. Crick recalled that neither felt compelled to spare the
other's feelings in this man-to-man exchange: when one posited a theory, the other sought to "demolish it in a candid but non-hostile manner." Neither feared miscues, short cuts, or dead ends—cocksureness
wouldn't let him. Emboldened by what each man had been led to believe about himself as an heir to brilliant men, they had both acquired
the "youthful arrogance" to think recklessly.[59]

Looking back, Crick could hardly blame Franklin for proceeding
with less abandon. As a woman, she had to be a careful, consummate
professional, rejecting rash hypotheses that her data could not yet confirm. Crick believed that what all good scientists needed was a partner
who communicated with "perfect candor, rudeness if need be." In the
Eighth Day of Creation, Horace Freeland Judson lamented that Franklin
had never had that partner at Kings: "It is evident from her notebooks
that she needed one, and clear from what we know of her character
that she would have worked well—candidly, rudely if need be—with
the right one." Her graduate student Raymond Gosling was the only
associate with whom she occasionally reviewed her findings; he, too,
agreed that she needed "someone of her own standing" with whom to
have frank discussions. She loved a good scientific argument and never
let it get personal, yet her collecting of data at King's remained a long
and lonely process.[60]

For Crick and Watson it was no process at all, since they carried out
no experiments of their own. Watson understood Franklin's data only
superficially, and Crick was frustrated that at the November talk in 1951
Watson didn't take notes, for he recalled inaccurately her highly technical jargon and measurements. Watson's misapprehending Franklin's
terms "unit cell" and "asymmetric unit" was responsible for the delay
in configuring their final model. Still, Watson was clever enough to
keep abreast of Franklin's progress and to pry details of her work out
of Wilkins whenever he would come round to complain about Franklin
into Watson's sympathetic ear. Watson knew that Franklin's impeccable
diffraction technique would ultimately yield the visual evidence to solve
the structure, for Wilkins was showing him images of hers that were
getting "prettier and prettier" by the day. She had photographs of two
DNA forms—A, or "dry," and B, or "wet"—and meticulous notes on the
amount of water in each. She concluded that the phosphates were lo-

cated on the outside of the molecule, encased in a shell of water, but the helical structure remained ambiguous. The B-form measurements supported helicalism, but calculations from her A-form images were inconclusive.[61]

Franklin spent a year reconciling the disparity, ultimately deciding that both forms consisted of two helices. Watson, meanwhile, proceeded as if helices were a fact. Early in 1953 Wilkins went into Franklin's lab and pulled out the image she had labeled "Photo 51," the "B form" of DNA. When he showed it to Watson, it made Watson's heart race; the pattern was simpler than anything he had seen before. Although he couldn't ascertain exact measurements, the crosses were formed so clearly in Franklin's image that Watson knew he was holding a portrait of a helix. He pressed Wilkins at dinner for more density values, water-content data—anything that could help him discern how many strands she was dealing with. Linus Pauling had figured three, and Wilkins's limited knowledge of the data did nothing to suggest otherwise. Crick and Watson scrapped their old models and began working on a new one based on the insights provided in Photo 51. They had no idea that Franklin was in the midst of figuring out mathematically what they would soon show with their model; and she was oblivious to the fact that her image had opened the floodgates.

Soon they were helped along by another windfall of her making: Max Perutz, an unsuspecting accomplice, had collected reports for the Medical Research Council (MRC) and saw no reason to deny Watson and Crick's request to see the one Franklin had submitted a year earlier, complete with measurements of the face-centered monoclinic unit cell. The density data supported the theory of two chains, as distinct from Pauling's hypothesis of three. Franklin had concluded that the B-form diffraction pattern appeared helical and likely contained two to four coaxial nucleic acid chains per helical turn.[62]

Crick and Watson's reading of the report has been compared to an enemy's happening to find the other side's codebook. The metaphor would be more appropriate had Franklin been aware of a war being waged. Within days of receiving the MRC report Watson and Crick disclosed their model of the double helix: two intertwined polynucleotide chains on which each adenine nucleotide corresponded to a thymine nucleotide, guanine to cytosine. Hydrogen bonds formed between the two opposite nucleotides at each rung of the molecule, making all of it consistent with Franklin's data and images. They drafted a paper for submission to *Nature*, despite having no data of their own. Watson's sister typed the text; Crick's wife drew the twisting chains of the helices. When Franklin heard they had figured out the base pairs she took a train to Cambridge to see the tinker toy model herself. She was gracious, even as Watson related it. He supposed that even she, in her stubbornness, could see that "the structure was too pretty not to be true." In the spirit of cooperation she offered a paper containing measurements and images, including the infamous photograph 51, thinking that no one had yet seen it.[63]

Franklin was likely so conciliatory because the model was no surprise; her supporting paper was a simple redraft of one she had already written. Wilkins and team submitted a third paper, "Molecular Structure of Deoxypentose Nucleic Acids," to round out the submissions to *Nature*. Crick and Watson's was the lead paper, the one that announced the discovery, while the others supported their claims. Nowhere in their submission did they include explanations for how they obtained their experimental proof, just a vague aside: "We have also been stimulated by a knowledge of the general nature of the unpublished experimental results and ideas of Dr. M. H. F. Wilkins, Dr. R. E. Franklin, and their co-workers at King's College, London." Crick apparently offered to put all three papers under one title so that the King's people would be coauthors, but Wilkins refused. Franklin was not presented with an offer to consider; then and always, Crick and Watson let Wilkins act on her behalf.[64]

After the fact, microbiologist André Lwoff pondered other roads that might have been taken more honorably and honestly: Crick and Watson could have offered Franklin joint authorship directly or they could have submitted a paper strictly on the scheme of base pairing, since it required the use of less of her data. A year later Crick and Watson explained in the footnote of another paper that their formulations of DNA structure "would have been most unlikely" without King's data. Many colleagues thought the gesture was too little, too late.[65] Franklin had long applied for a transfer to Birkbeck College; letters to friends confirm that she felt too beaten down at King's to stay. Randall told her that she could not take her data with her, a mandate that he would never have passed down to men in the unit. His terse advice was to take up something else, which she did. Under J. D. Bernal she worked primarily on the structure of the tobacco mosaic virus. But before she left King's she had the grace to tie up ends, providing a full Patterson synthesis of the A form and confirming the DNA structure in crystallographic terms.[66]

The world recognized Crick and Watson as the discoverers of the double helix, but Crick admitted that it would be more accurate to say that the double helix got *them* discovered. Crick finished his dissertation, left for Brooklyn Polytechnic and a visiting professorship at Harvard, and won a Lasker Prize and fellowships from the American Academy of Science and the Salk Institute before winning his Nobel Prize. Watson, too, became a sought-after commodity in biological science as well as a featured bachelor in *Vogue* magazine. He returned to the United States to a senior research fellowship at Caltech and in 1956 joined the faculty of the biology department at Harvard. In 1968 he left to become the director of Cold Spring Harbor, six years after winning the Nobel Prize. He presented an acceptance speech in Stockholm alongside Crick and Wilkins. Only Wilkins mentioned Franklin that night, and only in a passing point of information.[67]

Debates persist over whether or not Rosalind Franklin should have won a posthumous prize. Crick conceded that she would have figured out the puzzle of DNA within months, if not weeks, of his discovery; all she had left to sort out was the significance of Chargaff's rule and the pairing of protein bases. Lynne Osman Elkin has looked over her notebooks and agreed that Franklin "was very close"; unlike anyone else, Franklin knew the parameters of the helical backbone of the structure. She was the one who determined that DNA had two forms and was aware of hydrogen bonding and the differences between the enol and keto forms. Had she simply published the original draft of the paper she submitted with Crick and Watson's at the time that she wrote it, they would have been forced to credit her for substantiating their claims.[68]

Watson doesn't refute this, but insists it took his blend of outside perspective and opportunism to put all the pieces of the puzzle in place. "We used her data to think about, not to steal," he said in 1984, and in the twenty-first century he has remained steadfast in his characterization of Franklin as anti–model building and anti helical. He saw her as a doer, not a thinker, an observer, not an analyst. Like the women of the Harvard Observatory, she had the patience for mundane data collection, and he the propensity for epiphany. Aaron Klug, the South African chemist who worked side by side with Franklin at Birkbeck, viewed Franklin's deliberateness differently. "It's not just good needlework," he countered. "She worked *beautifully*. . . . The kind of single-mindedness that she had made her an absolutely first-class experimental worker." Moreover, her notebooks proved that she had the analytical mind Watson revered, for privately she pointed to helicalism well before Watson thought of it himself.[69] Notes she wrote for the 1951 colloquium Watson attended, for example, acknowledged the presence of "a big helix or several chains." Her belief that the sugar-phosphate groups were on the *outside* of the molecule was hers alone for a long time, and only when Crick proposed taking her advice in this respect did he and Watson create a successful model of DNA. Her images of the A and B forms suggested a helical structure, but didn't prove it. "We are not going to speculate," she told Gosling; "we are going to wait." Such was the mark of a conscientious scientist, Klug believed; she wouldn't read into what wasn't there.[70]

If Franklin had ever been tempted to right the record herself, she lost her chance when ovarian cancer took her life in April 1958, her illness likely caused by radioactive exposure; Gosling had worried about her standing for long hours in front of X-ray beams, adjusting and re-adjusting her lenses. During the eighteen months in which she suffered with the disease she continued to work in the lab, stopping only for surgeries or during periods of unbearable pain. Had she lived past the age of thirty-seven, there is no telling where her work on the tobacco mosaic virus would have led. She had already determined the location of its RNA and moved on to other viruses that likely would have resulted in the winning of a Nobel Prize with Klug, who won one in 1982. Upon her death, colleagues at Birkbeck chose to remember her accomplishments, not her slights. Bernal wrote a heartfelt tribute for *Nature* and *The London Times*: "As a scientist Miss Franklin was distinguished by extreme clarity and perfection in everything she undertook. Her photographs are among the most beautiful X-ray photographs of any substance ever taken. . . . She did nearly all this work with her own hands. At the same time she proved to be an admirable director of a research team and inspired those who worked with her to reach the same high standard."[71]

Fixated on the DNA race at King's, most people will likely never know that Franklin enjoyed social camaraderie with men. Although Crick admitted to being one of her patronizers in 1951, he befriended her once she left for Birkbeck and took care of her at his home after her fatal diagnosis. It is interesting to note that Crick, Klug, Bernal, and others who ended up singing her praises were men whose married status nullified or sublimated the sexual tensions of the lab. Studies have since confirmed that men prefer mentoring and collaborating with married women, to avoid sexual tensions and misunderstandings. Crick noted Franklin's preference for married colleagues too; she befriended their wives, lavished affection on their children, and breathed relief to be among men who didn't conjecture wildly about her sex life—at least not within earshot of her or their wives.[72]

Many of the details of the DNA race we know because Watson chronicled them thirteen years later in a manuscript titled "Honest Jim," later renamed "Base Pairs" and eventually "The Double Helix." His dream was to write a book as good as *The Great Gatsby*. Editors agreed that it was absorbing, but Houghton Mifflin turned it down because of its potentially libelous material. Harvard University Press agreed to publish it but backed out when scientists began contesting facts in the book. Abbreviated versions were printed in *Atlantic Monthly*, and the complete form, *The Double Helix*, eventually became a best seller in 1968. Like a good tabloid, it was gossipy, loaded with sordid details about scientists' foibles. Watson's "behind the scenes" exposé crushed the Arrowsmith myth: scientists were ambitious and selfish and played games of approximation; their frailties tainted their science. Many of the faults Watson found in scientists were his own, but Pauling, Crick, Kendrew, and Wilkins charged that they had been gratuitously misrepresented in the volume and that Watson had done more than betray a professional trust—he had crossed lines of decency to glorify himself.[73]

Although *The Double Helix* is controversial, it is the version of events best known by elite scientists as well as suburban book club members, high school students, and popular editorialists. It has left an imprint on the American mind—not only of the crude sociology of science but also of Rosalind Franklin, the amorphous female presence in the lab whom Watson less than endearingly called "Rosy." In his hands, she was not a person, but a caricature—an old maid, an evil stepmother, a wicked witch—all combined in the form of a scientist. Watson sized up "Rosy" in sexual terms, but she emitted no sexuality back to endear him. "By choice she did not emphasize her feminine qualities," he explained. She wore no lipstick and her dress was drab. As she stood in front of an audience presenting her research, Watson could only wonder how she'd look if she vamped it up by taking off her glasses and styling her hair. Underneath her bluestocking exterior was an unstable and aggressive woman who grew enraged when colleagues challenged her. "She had a good brain," he conceded. "If she could only keep her emotions under control." A pivotal scene in *The Double Helix* was one in which "Rosy" took offense at his claim that DNA was helical. When he accused her of incompetence, she moved to strike him; Watson dodged her and walked away unscathed.[74]

Wilkins, who witnessed the scene, maintained that Watson exaggerated; Franklin defended herself with words, not fists. Max Perutz thought Watson had maligned a "gifted girl" for Watson's own aggrandizement, and Franklin's family thought the portrait so despicable that they preferred that she be forgotten altogether. Feminists, meanwhile, charged him with playing the sex card, using Franklin's emotionality and unwomanly appearance against her at the same time, if in fact his descriptions were accurate. A psychoanalyst might suggest that Franklin's unwillingness to be Watson's sexual object resulted in his own passive aggression, manifested in his defaming of her to the world. Mary Ellmann, a reviewer of *The Double Helix*, likely had it right when she diagnosed his hostility as a reaction to the sense of contradiction Franklin embodied. She was "the one bug in the helix," Ellmann explained, "the woman who *studies* DNA like a man. . . . Why couldn't she content herself with playing assistant to Wilkins (and over his shoulder, to Crick and Watson)? Why was she ambitious for herself as well as competent in X-ray diffraction photography? Why wouldn't she cooperate?"[75]

Apparently her contradictions no longer mattered once Watson had successfully used her data for his ends. In accounts of the aftermath of the DNA race, he described Franklin more appealingly as athletic (a "gutsy mountaineer"), sexy and sophisticated, not a prickly bluestocking but an "obsessively professional scientist," "direct and data-focused," who "would occasionally change out of her lab coat into an elegant evening gown and disappear into the night." In the days and months after his discovery, "Rosy" was converted into a warm, likable human being who saw the error in her ways. As she supplied the supporting data for his structure, Watson concluded that finally she produced "first-rate science, not the outpourings of a misguided feminist." John Lear, who reviewed *The Double Helix* for the *Saturday Review*, saw the irony in depicting her as the enlightened convert: "If Watson had been willing to consider Rosalind Franklin as an intellectual equal instead of deriding her as a mindless shrew, he could easily have seen how to accept her thesis that the sugar-phosphate backbone of the DNA structure must be on the *outside* and the plates *within*. That's the conclusion he reached in the end."[76]

Since 1968, feminists and filmmakers have offered correctives to Watson's depictions. Anne Sayre, Franklin's close friend, wrote a biography in which she charged Watson with inventing a new literary form: the "'as-if-it-were-true' memoir justified on the basis of 'I-didn't-know-better-then.'" "Rosy" was a figment of the imagination, someone Sayre couldn't recognize. People can rub people differently, Sayre conceded, but you cannot put glasses on a woman with "the eyesight of an eagle." "Those spectacles were not a matter of opinion," she chided, yet they served as the finishing touch to "Rosy's" dowdy costume, just as the label "Rosy" itself was a fictitious flourish for effect. Anyone who knew the woman, Sayre insisted, would not have dreamed of calling her anything but Rosalind.[77]

If his depictions were insensitive, Watson followed the advice of a female editor and attached qualifying statements about Franklin in the epilogue to *The Double Helix*:

> Since my initial impressions of her, both scientific and personal (as recorded in the early pages of this book), were often wrong, I want to say something here about her achievements. The X-ray work she did at King's is increasingly regarded as superb. The sorting out of the A and B forms, by itself, would have made her reputation; even better was her 1952 demonstration, using Patterson superposition methods, that the phosphate groups must be on the outside of the DNA molecule. . . . All traces of our early bickering were forgotten, and we [Watson and Crick] both came to appreciate greatly her personal honesty and generosity, realizing years too late the struggles that the intelligent woman faces to be accepted by a scientific world which often regards women as mere diversions from serious thinking.[78]

For all his enlightenment about sexism in science, Watson went on to write accounts of his early career that featured women in limited and stereotypical ways. In *Genes, Girls, and Gamow* (2001), he confessed to being a twenty-five-year-old more interested in sex than in science. In the transatlantic world he created, the women of note were mothers, wives, and sisters of the men he took seriously or the faceless objects of his sex-starved gaze. The occasional competent female in his world was a technician or typist. If an intellectually rigorous scientist in her own right—as in the case of Franklin and the British crystallographer Dorothy Wrinch—she was prickly, only "fun" once she conceded defeat in the game of discovery and returned to the social play of women. In his world of science and philandering, women functioned as fantasy material or as pawns that men placed into close proximity to male scientists and their data. Linus Pauling's daughter, John Kendrew's wife, and even Watson's own sister were women Watson exploited for his competitive designs. He described his double helix as a woman who finally revealed herself to him. She was "beautiful," "too elegant" to refute. She was, like all women, an object of adoration.[79]

So how does one reconcile the Watson who treated women like diversions (usually sexual ones) and the one who supplied the sensitive insights in the epilogue of 1968? No doubt pressures of the nascent feminist movement made his apology a necessity, but he defended his book as a rendition of not what *is*, but what *was*—as his scientific fraternity saw it in 1953. Men at King's really did feel that Rosalind had to be "put in her place"; it really was the prevailing sentiment that "the best home for a feminist was in another person's lab." Other Nobel-winning scientists saw their discoveries through sexual lenses. Richard Feynman told a room full of men at Caltech, for example, that he had fallen deeply in love with QED, and though, like any aging woman, she no longer turned heads, his theory deserved respect for mothering children who did. Young Watson's world was a milieu in which George Gamow convened the exclusive "RNA Club," in which members went by amino acid code names and wore neckties with double helixes stitched into them.[80] How could Franklin have fit into this nexus?

Watson warned young scientists in his 2007 memoir, *Avoid Boring People*, that "mopping up the details . . . will not likely mark you out as an important scientist. Better to leapfrog ahead of your peers." Although Rosalind Franklin had all the data to come up with the structure of DNA before he did, what she lacked, as he saw it, was his spark of impetuous energy. E. B. Wilson used the same logic in assessing the work of unsuspecting Nettie Stevens in the early 1900s; he and Stevens came up with identical conclusions about the mechanisms of sex identity, but he packaged their work collectively as her female labor clarified by his brilliant synthesis. His assumptions about women's limitations persist. A woman lamented to Vivian Gornick in the 1980s that her male associates still classified their peers into idea people, technique people, and people who work hard. "They think women fall into the last category most frequently, if not always."[81] This labeling is an integral part of the story of women prizewinners and nonprizewinners, as is the social organization of the science itself.

Notes

1. Willard Libby, Bernd Matthias, Lothar Nordheim, and Harold Urey, "Maria Goeppert Mayer, Professor of Physics, 1906–1972," Box 9, folder 8, Maria Goeppert Mayer Papers, Special Collections, University of California at San Diego (hereafter MGM).

2. Edward Teller with Judith L. Shoolery, Memoirs: A Twentieth-Century Journey in Science and Politics (Cambridge, MA: Perseus, 2001), 241–42.

3. Maria Goeppert Mayer, "The Shell Model," Nobel Lecture, Nobel Lectures, Physics, 1963–1970 (Amsterdam: Elselvier, 1972), 29–30.

4. Joan Dash, A Life of One's Own: Three Gifted Women and the Men They Married (New York: Paragon House, 1988), 311.

5. Horace Freeland Judson, "No Nobel Prize for Whining," New York Times, October 20, 2003; Noemie Bencer-Koller, "Chien-Shiung Wu," in Out of the Shadows: Contributions of Twentieth-Century Women to Physics, ed. Nina Byers and Gary Williams (New York: Cambridge University Press, 2006), 260–66.

6. C. S. Wu, E. Ambler, R. W. Hayward, D. D. Hoppes, and R. P. Hudson, "Experimental Test of Parity Conservation in Beta Decay," Physical Review 105 (1957): 1413; L. M. Jones, "Intellectual Contributions of Women to Physics," in Women of Science: Righting the Record, ed. G. Kass Simon and Patricia Farnes (Bloomington: Indiana University Press, 1990), 205–8.

7. Sharon Bertsch McGrayne, Nobel Prize Women in Science (Washington, DC: Joseph Henry Press, 1998), 268, 276.

8. Daniel Kevles, The Physicists: The History of a Scientific Community in Modern America (Cambridge, MA: Harvard University Press, 1987), 227; Sandra Harding, The Science Question in Feminism (Ithaca, NY: Cornell University Press, 1986), 76.

9. Ferenc Morton Szasz, The Day the Sun Rose Twice: The Story of the Trinity Site Nuclear Explosion, July 16, 1945 (Albuquerque: University of New Mexico Press, 1984), 9–10; Leona Marshall Libby, The Uranium People (New York: Charles Scribners' Sons, 1979), 45–51; Ruth Lewin Sime, Lise Meitner: A Life in Physics (Berkeley: University of California Press, 1996), 326–27, 369–70; "Lise Meitner," in Out of the Shadows, 74–82; Jones, "Intellectual Contributions of Women to Physics," 193–95.

10. Helena M. Pycior, Nancy G. Slack, and Pnina G. Abir-Am, eds., Creative Couples in the Sciences (New Brunswick, NJ: Rutgers University Press, 1996), ix, 3; Ferdinand V. Coroniti and Gary A. Williams, "Susan Jocelyn Bell Burnell," in Out of the Shadows, 419–26.

11. Stephen G. Brush, "Nettie M. Stevens and the Discovery of Sex Determination by Chromosomes," in Sally Gregory Kohlstedt, ed., History of Women in the Sciences:

Readings from Isis (Chicago: University of Chicago Press, 1999), 343–44; G. Kass Simon, "Biology Is Destiny," in *Women of Science*, 225–26.

12. Harriet Zuckerman, *Scientific Elite: Nobel Laureates in the United States* (New York: The Free Press, 1977), 53–54; Marianne Wenzel quoted in McGrayne, *Nobel Prize Women*, 189.

13. Maria Mayer to Hans Jensen, [November 1949]; Jensen to Mayer, November 2, 1949; Jensen to Mayer, December 22, 1949, Box 1, folder 16; "The Moon, the Atom . . . ," 10–11, Box 9, folder 24, MGM; Maria Goeppert Mayer and J. Hans Jensen, *Elementary Theory of Nuclear Shell Structure* (New York: Wiley and Sons, 1955).

14. C. W. Wong, "Harriet Brooks," in *Out of the Shadows*, 66–73; Ruth Howes and Caroline Herzenberg, *Their Day in the Sun: Women of the Manhattan Project* (Philadelphia: Temple University Press, 1999), 36; Londa Schiebinger, *Has Feminism Changed Science?* (Cambridge, MA: Harvard University Press, 1999), 95–96; Natalie Angier, "Geek Chic and Obama, New Hope for Lifting Women in Science," *New York Times*, January 20, 2009, science section, 1, 4.

15. It is interesting to note that single women did not start winning Nobel Prizes in science until after the women's movement began in the late 1960s and early 1970s. After Maria Mayer (1963), Dorothy Crowfoot Hodgkin (1964), and Rosalyn Yalow (1977), the next Nobelists were the single women Barbara McClintock (1982), Rita Levi-Montalcini (1986), and Gertrude Elion (1988). Margaret Rossiter, *Women Scientists in America: Struggles and Strategies to 1940* (Baltimore: Johns Hopkins University Press, 1982), 393; *Women Scientists in America: Before Affirmative Action, 1940–1972* (Baltimore: Johns Hopkins University Press, 1995), 329–32; Pnina Abir-Am, forward to *Creative Couples*, xi; Schiebinger, *Has Feminism Changed Science?* 97.

16. Maria Mayer to Lisa Keller, March 17, 1969, Box 2, folder 9; "The Moon, the Atom . . . ," 7; "Maria Goeppert Mayer," typed autobiography, Box 9, folder 2, MGM.

17. See, for example, Elizabeth Rona, *How It Came About: Radioactivity, Nuclear Physics, Atomic Energy* (Oak Ridge, TN: Oak Ridge Associated Universities, 1978), 1; Evelyn Fox Keller, *A Feeling for the Organism: The Life and Work of Barbara McClintock* (New York: Henry Holt, 1983), 20–22, 34–35; Rita Levi-Montalcini, *In Praise of Imperfection: My Life and Work* (New York: Basic Books, 1988), 4–16; Fay Ajzenberg-Selove, *A Matter of Choices: Memoirs of a Female Physicist* (New Brunswick, NJ: Rutgers University Press, 1994), 3, 14, 20; Moira Reynolds, *American Women Scientists: 23 Inspiring Biographies, 1900–2000* (Jefferson, NC: McFarland, 1999), 112.

18. Alice Rossi, "Barriers to the Career Choice of Engineering, Medicine, or Science Among American Women," in *Women and the Scientific Professions: The MIT Symposium on American Women in Science and Engineering*, ed. Jacquelyn A. Mattfeld and Carol G. Van Aken (Cambridge, MA: MIT Press, 1965), 91; Anne Roe, *The Making of a Scientist* (New York: Dodd, Mead, 1953), 92; Philip Wylie, *Generation of Vipers* (New York: Rinehart, 1942).

19. Zuckerman, *Scientific Elite*, 66; E. N. Plank and R. Plank, "Emotional Components in Arithmetic Learning as Seen Through Autobiographies," in *The Psychoanalytic Study of the Child* (New York: International University Press, 1954); Evelyn Fox Keller, *Reflections on Gender in Science* (New Haven, CT: Yale University Press, 1995), 91n; Londa Schiebinger, *Has Feminism Changed Science?* 59; Joan Dash, *Triumph of Discovery: Women Scientists Who Won the Nobel Prize* (Englewood Cliffs, NJ: Julian Messner, 1991), 3, 233–40; Olga Opfell, *The Lady Laureates: Women Who Have Won the Nobel Prize* (Metuchen, NJ: Scarecrow Press, 1986), 226–27; Reynolds, *American Women Scientists*, 81; McGrayne, *Nobel Prize Women*, 175–80.

20. Mary Harrington Hall, "The Nobel Genius," 108, Box 9, folder 30, MGM.

21. Transcript of tape-recorded interview of Maria Goeppert Mayer by Thomas Kuhn, February 20, 1962, tape 2, side 1, Center for History of Physics, American Institute of Physics, College Park, MD; "The Moon, the Atom . . . ," 7.

22. Interview with Maria Goeppert Mayer, February 20, 1962; McGrayne, *Nobel Prize Women*, 180–82; Dash, *A Life of One's Own*, 244, 253.

23. Transcript of interview between Joseph Mayer and Lillian Hoddeson, January 24, 1975, 13–14, Center for History of Physics, American Institute of Physics; McGrayne, *Nobel Prize Women*, 182.

24. "The Moon, the Atom . . . ," 8; Maria Mayer to Joan Hellweg, March 29, 1968, Box 2, folder 8; Hall, "The Nobel Genius," 69; "Biography," Box 9, folder 2, MGM; McGrayne, *Nobel Prize Women*, 183–84; Jones, "Intellectual Contributions of Women to Physics," 200.

25. Jones, "Intellectual Contributions of Women to Physics," 189–91; Helmut Rechenberg, "Hertha Spöner," 127–36; Peggy Aldrich Kidwell, "Margaret Eliza Maltby," 26–35, in *Out of the Shadows*.

26. Rossiter, *Women Scientists in America: Before Affirmative Action*, 122–64; McGrayne, *Nobel Prize Women*, 184; Barbara Shiels, *Winners: Women and the Nobel Prize* (Minneapolis: Dillon Press, 1985), 98; Dash, *Triumph of Discovery*, 8–12.

27. Maria Goeppert Mayer and Karl Herzfeld, "On the Theory of Dispersion," *Physical Review* 49 (1936): 332; Maria Mayer and Alfred Sklar, "Calculations of the Lower Excited Levels of Benzene," *Journal of Chemical Physics* 6 (1938): 645; Maria Goeppert Mayer and Robert G. Sacks, "Calculations on a New Neutron-Proton Potential," *Physical Review* 53 (1938): 991; "The Moon, the Atom . . . ," 8; Interview with Maria Goeppert Mayer, February 20, 1962; Steven A. Moszkowski, "Maria Goeppert Mayer," in *Out of the Shadows*, 208.

28. Alfred Sklar to Mrs. Mayer, March 4, 1944, Box 1, folder 11, MGM.

29. Maria Goeppert Mayer, "Double Beta-Disintegration," *Physical Review* 48 (1935): 512; Interview with Joseph Mayer, January 24, 1975, 6; McGrayne, *Nobel Prize Women*, 187; Libby, Matthias, Nordheim, and Urey, "Maria Goeppert Mayer"; "Maria Goeppert Mayer," typed autobiography.

30. Transcript of tape-recorded interview of Edward Teller by Karen Fleckenstein, September 9, 1983, Center for the History of Physics, American Institute of Physics, 1–2; Hall, "Nobel Genius," n.p.; Dash, *Life of One's Own*, 276–79.

31. Hall, "Nobel Genius," n.p.; Dash, *Triumph of Discovery*, 12; Dash, *Life of One's Own*, 282, 288–89; Reynolds, *American Women Scientists*, 83–84; Opfell, *Lady Laureates*, 230.

32. Max Born to Maria Mayer, June 8, 1939, Box 1, folder 6; John Kirkwood to Mrs. Mayer, January 9, 1941, Box 1, folder 8, MGM; Hall, "Nobel Genius," 108; Shiels, *Winners*, 101; Darlene Stille, *Extraordinary Women Scientists* (Chicago: Children's Press, 1995), 125; Reynolds, *American Women Scientists*, 84.

33. Laura Fermi, *Atoms in the Family: My Life with Enrico Fermi* (Chicago: University of Chicago Press, 1954), 80–81; Hall, "Nobel Genius," 68, 108; Dash, *Life of One's Own*, 301.

34. Constance Warren to "Mrs. Mayer," April 18, 1942; April 23, 1942, Box 1, folder 9; Max Born to Maria Mayer, January 20, 1943, Box 1, folder 10; Maria Mayer to Constance Warren, January 16, 1945, Box 1, folder 12; Maria Mayer's speech for "Women in Science Program," 1964, Box 2, folder 3, MGM.

35. McGrayne, *Nobel Prize Women*, 192; Shiels, *Winners*, 104–105; Stille, *Extraordinary Women*, 125; Reynolds, *American Women Scientists*, 84; Dash, *Triumph of Discovery*, 16–17.

36. Dash, *A Life of One's Own*, 300–301; *Triumph of Discovery*, 17–20; Opfell, *The Lady Laureates*, 231–32; Hall, "Nobel Genius," 68; The University of Chicago Comptroller to

Maria Mayer, September 17, 1958, Box 1, folder 25, MGM; Schiebinger, *Has Feminism Changed Science?* 59.

37. Hall, "Nobel Genius," 110; Dash, *Life of One's Own*, 314; Shiels, *Winners*, 105; Joe Mayer to Peter Mayer, September 24, 1953; Maria Goeppert Mayer to "Toots," September 25, 1953, Box 1, folder 20; Mary Markley, "Maria Mayer: Physicist," *Christian Science Monitor*, July 17, 1964, Box 9, folder 30; "Maria Goeppert Mayer," typed autobiography, 2–3, MGM.

38. Dash, *Triumph of Discovery*, 27–28; *Life of One's Own*, 321; Opfell, *Lady Laureates*, 235; Teller, *Memoirs*, 241–42; Jones, "Intellectual Contributions of Women to Physics," 237.

39. Within months of Mayer's first contract, Roger Revelle changed the terms so that she was paid full time for nine months ($14,208 instead of the original $6,780). Roger Revelle to Joe and Maria Mayer, July 14, 1959; October 9, 1959, Box 1, folder 26; "The Moon, the Atom . . . ," 2, MGM; Rossiter, *Women Scientists in America: Before Affirmative Action*, 162.

40. "Pre-Dawn Party (For 2)—That's Nobel Prize Winner's Celebration," *San Diego Union*, Box 9, folder 28, MGM.

41. Linda McCausland to Dr. Mayer, November 10, 1963; Stella Hull to Maria Mayer, February 13, 1964; Marcia George to Maria Mayer, December 7, 1963; Maria Mayer to Marcia George, January 2, 1964, Box 10, folder 19, MGM.

42. Burton Feldman, *The Nobel Prize: A History of Genius, Controversy, and Prestige* (New York: Arcade, 2000), 1–23; Zuckerman, *Scientific Elite*, 19–20.

43. Zuckerman, *Scientific Elite*, 9–10; E. L. Tatum to Maria Mayer, April 15, 1965; Tatum to U.S. Nobel Laureates in Science, July 7, 1965; H. G. Franck to Dear Sir, July, 1965; Box 2, folder 5, MGM.

44. Carol Kahn, "She Cooks, She Cleans, She Wins the Nobel Prize," *Family Health* 10 (June 1978): 24–27; "The Chemistry-Minded Mother," *Time*, November 6, 1964, 41; Eileen Keerdoja and William Slate, "A Nobel Woman's Hectic Pace," *Newsweek*, October 29, 1979; "British Winner Is a Grandmother," *New York Times*, October 30, 1964, 23–24; Gina Maranto, "At Long Last—a Nobel for a Loner," *Discoverer*, December 1983, 26; Leticia Kent, "Winner Woman," *Vogue*, January 1978, 131; Mary Harrington Hall, "An American Mother and the Nobel Prize—a Cinderella Story in Science," *McCall's*, July 1964.

45. Feldman, *The Nobel Prize*, 1; Kevles, *The Physicists*, 197; Jonathan R. Cole, *Fair Science: Women in the Scientific Community* (New York: The Free Press, 1979), 8; Robert K. Merton, "The Matthew Effect in Science," *Science* 199 (January 5, 1968): 55–63; Margaret Rossiter, "The Matilda Effect in Science," *Social Studies of Science* 23, no. 2 (1993): 325–41.

46. Zuckerman, *Scientific Elite*, 96–100.

47. Mildred Cohn, "Carl and Gerty Cori, A Personal Recollection," in *Creative Couples*, 72–75, 82; Interview of Dr. David Kipnis in "Living St. Louis," KETC St. Louis, 2002; McGrayne, *Nobel Prize Women*, 93.

48. Transcript of Interview with Carl Cori, October 18, 1982, Department of Biological Chemistry, Harvard Medical School, Washington University School of Medicine Oral History Project, Bernard Becker Medical Library, St. Louis, MO; Cohn, "Carl and Gerty Cori," 77, 81; Opfell, *The Lady Laureates*, 213; McGrayne, *Nobel Prize Women*, 105–6, 113; Interview of David Kipnis.

49. Zuckerman, *Scientific Elite*, 99–100.

50. Charlie Rose interview with James Watson, *The Charlie Rose Show*, Public Broadcasting Service, June 10, 2003.

51. Brenda Maddox, *The Dark Lady of DNA* (New York: Harper Collins, 2002), xvii–xviii.

52. Anne Sayre, *Rosalind Franklin and DNA* (New York: W. W. Norton, 1975), 38–63.

53. Sayre, *Rosalind Franklin*, 59–79.

54. Maureen M. Julian, "Women in Crystallography," in *Women of Science*, 335.

55. Horace Freeland Judson, *Eighth Day of Creation: The Makers of the Revolution in Biology* (New York: Simon and Schuster, 1979), 101–4; James D. Watson, *Avoid Boring People: Lessons from a Life in Science* (New York: Knopf, 2007), 98.

56. Judson, *Eighth Day of Creation*, 136, 141, 148; Maddox, *Dark Lady of DNA*, 127–34, 137, 160–61; Sayre, *Rosalind Franklin*, 84–105; Edward Edelson, *Francis Crick and James Watson and the Building Blocks of Life* (New York: Oxford University Press, 1998), 44–45.

57. Watson, *Avoid Boring People*, 38–93; Errol C. Friedberg, *The Writing Life of James D. Watson* (Cold Spring Harbor, NY: Cold Spring Harbor Laboratory Press, 2005), 12; Cole, *Fair Science*, 8, 132–33.

58. Francis Crick, *What Mad Pursuit: A Personal View of Scientific Discovery* (New York: Basic Books, 1988), 4, 23, 45, 64; Friedberg, *Writing Life of Watson*, 5–8; Edelson, *Francis Crick and James Watson*, 16–25; Gunther S. Stent, introduction to *The Double Helix: A Personal Account of the Discovery of the Structure of DNA*, by James D. Watson, ed.Gunther S. Stent (New York: W. W. Norton, 1980), xvii; Watson, *Avoid Boring People*, 95–99.

59. William F. Bragg, "Preface," 1; James D. Watson, *The Double Helix*, 7–14; Judson, *Eighth Day of Creation*, 128; Crick, *What Mad Pursuit*, 60–70; Edelson, *Francis Crick and James Watson*, 24–25.

60. Judson, *Eighth Day of Creation*, 147, 149.

61. Watson, *Double Helix*, 48, 86; Crick, *What Mad Pursuit*, 65; Judson, *Eighth Day of Creation*, 121–23.

62. Watson, *Double Helix*, 98, 104–5; Lynne Osman Elkin interviewed on *Nova* online, http://www.pbs.org/wgbh/nova/photo51/elkin/html, September 4, 2008; Judson, *Eighth Day of Creation*, 160–61.

63. Watson, *Double Helix*, 122–24; James D. Watson, *Genes, Girls, and Gamow: After the Double Helix* (New York: Vintage Books, 2001), 11.

64. J. D. Watson and F. H. C. Crick, "Molecular Structure of Nucleic Acids," *Nature*, April 25, 1953, 737; James D. Watson with Andrew Berry, *DNA: The Secret of Life* (New York: Alfred A. Knopf, 2003), 55; Sayre, *Rosalind Franklin*, 158–61; Charlie Rose interview, *Charlie Rose Show*; Watson, *Avoid Boring People*, 107.

65. Gunther S. Stent, "A Review of the Reviews," in *Double Helix*, 161–75 (originally in *Quarterly Review of Biology* 43, no. 2 (1968): 179–84; Linus Pauling, "Molecular Basis of Biological Specificity," *Nature*, April 26, 1974, 769–71; Sayre, *Rosalind Franklin*, 192, 210; Maddox, *Dark Lady of DNA*, 210.

66. J. T. Randall to R. E. Franklin, April 17, 1953, Papers of Rosalind Franklin, Churchill Archives Center, Churchill College, Cambridge, U.K. (letter digitized in cooperation with the U.S. National Library of Medicine, Digital Manuscripts Project, Bethesda, MD); Judson, *Eighth Day of Creation*, 186–87.

67. Crick, *What Mad Pursuit*, 76; Watson, *Girls, Genes, and Gamow*, 98–99; Friedberg, *Writing Life of Watson*, 14.

68. Francis Crick, "The Double Helix: A Personal View," *Nature*, April 26, 1974, 766–71; *What Mad Pursuit*, 75; Lynne Osman Elkin interview, *Nova* online.

69. Maddox, *Dark Lady of DNA*, 315; John Rennie, "A Conversation with James D. Watson," *Scientific American* 288 (April 2003): 66; Watson with Berry, *DNA*, 50, 55; Watson, *Genes, Girls, and Gamow*, 10; Charlie Rose Interview, *Charlie Rose Show*; Watson, *Avoid Boring People*, 105.

70. Aaron Klug, "Rosalind Franklin and the Discovery of the Structure of DNA," *Nature*, August 24, 1968, 43–44; Judson, *Eighth Day of Creation*, 102–3, 118–20, 125, 128, 149, 159; A. Klug to Dr. P. Siekevitz, April 14, 1976, Rosalind Franklin Papers.

71. "Dr. Rosalind Franklin," *The London Times*, April 19, 1958. Bernal's tribute taken from Sayre, *Rosalind Franklin*, 180.

72. J. Scott Long, "The Origins of Sex Differences in Science," *Social Forces* 68 (1990); "Productivity and Academic Position in the Scientific Career," *American Sociological Review* 43 (1978); Crick, *What Mad Pursuit*, 86–87; Judson, *Eighth Day of Creation*, 149.

73. Stent, introduction, xxiv–xxv; Richard C. Lewontin, "'Honest Jim' Watson's Big Thriller about DNA," *Chicago Sunday Sun-Times*, February 25, 1968, 1–2; Crick, *What Mad Pursuit*, 80–81; Sayre, *Rosalind Franklin*, 212; Friedberg, *Writing Life of Watson*, 18–48.

74. Watson, *Double Helix*, 14–15, 45, 96.

75. Edelson, *Crick and Watson*, 58–60; Sayre, *Rosalind Franklin*, 129; Judson, *Eighth Day of Creation*, 159; Maddox, *Dark Lady of DNA*, 311–12; Mary Ellmann, "The Scientist Tells," *Yale Review* 57 (Summer 1968): 631–35.

76. Watson with Berry, *DNA*, 46; Watson, *Double Helix*, 122–24; John Lear, "Heredity Transactions," *Saturday Review*, March 16, 1968, 36, 86.

77. Wilkins admitted to using the name "Rosy," as did other colleagues behind Franklin's back. Judson, *Eighth Day of Creation*, 148; Sayre, *Rosalind Franklin*, 18–21, 191; Crick, *What Mad Pursuit*, 82.

78. Watson, *Double Helix*, 132.

79. Watson, *Genes, Girls, and Gamow*, 10–11, 18, 37, 42, 44, 85, 152, 170, 179, 252.

80. Richard Feynman, *The Feynman Lectures in Physics* (Reading, MA: Addison-Wesley, 1964), cited in Harding, *Science Question in Feminism*, 120; Watson, *Double Helix*, 14–15; Judson, *Eighth Day of Creation*, 265.

81. Watson, *Avoid Boring People*, 113; G. Kass Simon, "Biology Is Destiny," in *Women of Science*, 226; Vivian Gornick, *Women in Science: Then and Now* (New York: The Feminist Press, 2009), 66.

III AMERICAN WOMEN AND SCIENCE IN TRANSITION, 1962–

NAOMI WEISSTEIN HAD DREAMED OF A CAREER IN SCIENCE EVER since reading *Microbe Hunters* as a child. It's ironic that her inspiration was also that of James Watson, for their career paths diverged widely. She, too, had performed stunningly in college; the women who had mentored her were brilliant but taught at Wellesley because the research institutions that had trained them wouldn't hire them. They encouraged Weisstein to go to Harvard for graduate study, but there the problems began. Although graduate mentors encouraged Watson, deans cautioned Weisstein against competing with men in her class. One of the "star" professors told her directly that she didn't belong, and male students puffed on their pipes in agreement. No one at Harvard would give her access to equipment to complete her research, so she did it at Yale and returned to Harvard to collect her degree.

Weisstein got her PhD in psychology in 1964, in the aftermath of Maria Mayer's Nobel Prize and just as crystallographer Dorothy Crowfoot Hodgkin was winning hers. Nevertheless, their celebrity and Weisstein's top ranking in her class didn't help her to get a job. Administrators assumed that someone else had done her research, and they couldn't imagine her lecturing to intimidating numbers of men. She took a temporary lectureship at the university where her husband taught history; the money was meager, and she couldn't borrow books from the library. She was professionally isolated, but managed to run experiments anyway. When she submitted her research for publication, an editor who rejected it asked brazenly if he could verify her data and publish it for himself. If she remained unsure whether her sex was to blame, science conferences brought the sexism into sharp relief. Often she sat through slide presentations only to see images of bikinied bombshells inadvertently appear on the screen, accompanied by raucous laughter. Weisstein came to the conclusion that "making jokes at women's bodies constitute[d] a primary social-sexual assault." She realized that her professional experience mirrored women's experience in society at large. Changing men's science required changing men's world.[1]

For any woman coming of age in the 1950s and securing her first professional position in the early 1960s, the shift from embracing postwar gender roles to acquiring feminist sensibilities may have seemed impossibly difficult, but the seeds of fresh vision simply needed catalysts to germinate. In the case of women scientists, those catalysts were social, cultural, political, and economic, but they also included paradigmatic shifts within science itself.

After World War II Anne Roe's small cadre of elite men had stood as examples of scientific heroism, and Americans assumed by extension that all scientists were brilliant and eccentric to some degree—"masculine minded," as they defined it. But other social scientists were beginning to note a dissonance between the images of the scientist conjured by professional peers and sci-fi writers in the 1950s and the traits of people in real-world laboratories. A study conducted in 1947 and again in 1963 confirmed that the dissonance was intensifying. While Americans still perceived nuclear physics as a prestigious field, only 2 percent of them could describe the daily activities of physicists.[2]

Sociologist Bernice Eiduson conducted a study of forty Caucasian male scientists, most of whom were married with children, and published her findings as *Scientists: Their Psychological World* in 1962. Less interested in the "elite" than in the "everyday" practitioners of science, the ones who performed for paychecks rather than Nobel Prizes, she also introduced a new breed of scientific maverick whom she called the "gentleman scientist." Indeed he could be notoriously neurotic, prone to regression into infantile and other antisocial behavior, all tolerated because of his talent, but he also had a sense of decorum about him that had eluded Roe's esteemed but socially inept scientific elite. Like his predecessor, he was intelligent, but aspired to be cosmopolitan and multifaceted. He read the daily newspaper, traveled the world, knew fine cuisine—he was a Renaissance man of sorts, and he didn't let his science define him. Six years later, Walter Hirsch echoed Eiduson's claims that the real American scientist was more domestic and leisured than Roe's image. Eighty-six percent contributed to their community fund, and most attended plays, concerts, or museums as patrons of the arts.

Many of the traits Roe had idealized in her scientific elite Eiduson now deemed dysfunctional in modern collaborative environments. Once an autonomous maverick, the PhD scientist had become a technician, a mere "cog in the wheel. . . . Neither he nor any of his 'spokesmates' know where the vehicle is driving [to], nor why, nor exactly where his skills or contributions fit [in]," she determined. In such Taylorized environments, men were better off conceding their individual impulses. As Eiduson concluded, "The most valuable scientific man is not the thoughtful intellectual of 'old science,'" but rather the "man who recognizes and accepts the fact that neither he nor any man can perform the new technical job alone."[3]

Collaboration in science had been the reality before Eiduson declared it so, and by 1972 almost two-thirds of Nobel Prize winners had participated in joint discoveries, even as press coverage perpetuated their identities as "lonely geniuses." In 1953 Roe's future scientists were boys who had declined invitations to their sisters' tea parties in order to be alone with their thoughts; ten years later sociologists recommended that they pull chairs up to the table. Team players with interpersonal skills—the kinds who played well with others—were the ones more likely to succeed in Big Science. Unwittingly, a new idealized scientist had been created, one with social skills and a breadth of interests that were culturally female.[4]

Popular perceptions and cultural expectations of scientists changed gradually through the 1960s. A survey of college students similar to Mead and Metreaux's in the 1950s confirmed that, even when Roe's heroic scientist was respected for his dedication to science, boys didn't want to be him and girls didn't want to date him. If forced to choose the most "normal" of scientists, respondents picked biologists over physicists, perceiving the former to be more human and the latter too eccentric to admire. Roe had thought the disinterestedness of the scientific elite was admirable; from her perspective, no political or military man could quash Cold War anxieties as certainly as the objective men who had built the bomb. Such veneration waned a decade later, however. Walter Hirsch confirmed in 1958 that science fiction writers were painting more complicated portraits of their scientific "supermen" and "villains." They were becoming "real human beings . . . facing moral dilemmas, . . . recogniz[ing] that science alone is an inadequate guide for the choices they must make."[5]

Hirsch may have identified what Americans were beginning to think about scientists, but at the same time educators, government propaganda, and technical recruiters told the public that the acumen of the nation's scientists was a gauge of supremacy in international affairs. The "Sputnik effect" was an urgent response to scoring second technologically to the Soviets. The Kennedy administration pumped more money into the Department of Defense and the National Aeronautics and Space Administration (NASA) in the hope of surpassing the Soviets as the most technologically advanced nation in the world. By 1960 eight billion dollars of the national budget went to scientific research and development; that total increased to $13.6 billion just three years later.[5] The use of popular images and federal funds was supposed to make Americans want to become scientists, but there was also mounting cynicism about the omniscience and moral fiber of elite men of science. Ralph Lapp, an atomic physicist, conceded that the huge size of collaborative science made him dubious, for it was too easy for accountability to get lost in the thousands of scientists and millions of dollars modern projects required. Under Lapp's critical light the collaborative enterprise known as the Manhattan Project lost luster. Mary Palevsky, a child born of a love affair between Los Alamos physicists, wondered, too, how science could bring two people together and wreak havoc on humanity at the same time.[7]

Public disenchantment with Manhattan Project heroes shouldn't have been surprising, since many of them had become disenchanted themselves. As early as November 1947 Oppenheimer told a room of aspiring scientists at MIT that "physicists have known sin; and this is a knowledge which they cannot lose." Leo Szilard, once instrumental in convincing FDR to build the bomb, founded the Council for a Livable World in 1962 to prevent the misappropriation of nuclear arms. Physicists knew sin, but eventually scientists in all fields felt the weight of their moral dilemmas. The Federation of American Scientists was composed of members committed to analyzing the relationship of science to public policy. By 1969 Scientists for Social and Political Action sought "a radical redirection" for modern science altogether.[8]

Women who had helped make the bomb felt similar moral burdens, though they hadn't the pedestal from which to make amends. Los Alamos physicist Joan Hinton lobbied for atomic energy to be internationalized immediately after the war. As her former colleagues racked up publications and Nobel Prizes, she left physics and American science altogether to make carts in an iron factory in rural China. More than scientific research, Mao's Cultural Revolution appealed to her sense of humanity: "The Chinese with their bare hands are building up a new nation," she explained, "while the Americans . . . are preparing to destroy mankind." Science wives also felt the burden of having unquestioningly sacrificed their husbands to the Tech Area of Los Alamos. Feigned or real, their ignorance no longer felt excusable in the skeptical climate of the 1960s. Laura Fermi spoke out against nuclear proliferation, while Alice Kimball Smith wrote *A Peril and a Hope*, a history of the atomic scientist movement stripped of heroic mythology.[9] Roe had contended that the scientific elite's withdrawal from politics, religion, domestic, and emotional matters had made all of them impervious to influences that tainted judgment. Although Smith's husband had been one of Roe's sixty-four elite, she expressed a preference for the scientist who thoughtfully drew from the well of social experience.

Smith and Eiduson were scientific outsiders who evaluated scientists within their larger social and cultural contexts. The same relativist impulse also infiltrated science itself. Thomas Kuhn was the quintessential male type Roe had envisioned: he had studied theoretical physics at Harvard in the 1940s and had worked at Berkeley in the 1950s, Princeton in the 1960s, and MIT in the 1970s. Rather than pursue "pure" science solely, however, he ventured into scientific philosophy—a path taken at the same time by another Harvard physicist, Evelyn Fox Keller, who also would dismantle assumptions about the universality of science. The same year that Eiduson published her study Kuhn published his classic work, *The Structure of Scientific Revolutions*, in which he challenged the positivist tradition that had long enhanced the hegemony of masculine scientists. Scientific knowledge, as he viewed it, did not naturally accumulate over time to lead to greater truths. There were periods of equilibrium, punctuated by "intellectually violent revolutions," in which "normal science" as all knew it was rudely overturned. During this process, dissenting opinions from the margins gained the momentum to overthrow orthodox beliefs about the workings of the natural world. These shifts in paradigm revealed science and scientists as social entities shaped by their historical contexts. Simply put, men could not come to truths that they were not conditioned to tell.[10]

Kuhn concluded that scientific method was not value neutral, nor was the seemingly omniscient scientist. Philosophers Louis Althusser, André Gorz, and Jürgen Habermas all agreed that a dose of skepticism should be applied to any assessment of scientists and their endeavors. In *One-Dimensional Man* (1964) Herbert Marcuse wondered whether Western scientists weren't agents of the modern industrial machine. Antimilitary sentiment spawned by the Vietnam War further damaged the reputations of scientists, and physicists in particular, since they had cultivated strong ties to government money. By 1975 Paul Feyerabend condemned the out-and-out "chauvinism of science": "Science is neither a single tradition, nor the best tradition there is," he raged. "Neither science nor rationality are universal measures of excellence. They are particular traditions, unaware of their historical grounding."[11]

Kuhn's work on scientific revolutions had spawned its own revolution in the study of science as culture. None of it was explicitly feminist, yet all of it examined science's relationship to authority and power in ways feminists found helpful. One woman recalled her amusement as "androcentric scientists" read Kuhn's work and were left scratching their heads: what did Kuhn really mean when he declared that "proponents of competing paradigms practice[d] their trades in different worlds?" By virtue of their becoming scientists, she and her female colleagues understood completely what he meant about being torn in two. Although Feyerabend never identified the power relations of science as patriarchal, his ideas could also easily be used to describe women's struggles in the lab. Forces constituting "science" and "nonscience" he understood to be political; by extension, so were those that defined women as nonscientists.[12]

The very year Kuhn published *Scientific Revolutions*, wider debate ensued over a book called *Silent Spring* by the marine biologist Rachel Carson. The idea for the book was hatched when a friend wrote Carson lamenting the disappearance of birds in her hometown in Massachusetts; the friend wondered if the spraying of DDT was to blame, and Carson, too, had her suspicions in light of recent unchecked chemical use. The wonder drug thalidomide had been pulled from shelves after proving to cause birth defects in the newborns of women who took it, and in the case of DDT, producers ignored mounting evidence of its dangers for material gain. Men did not have a firm grasp of the human toll of the drugs or bombs or chemicals they synthesized in the lab, Carson feared, and yet they arrogantly unleashed them on the world.[13]

Immediately *Silent Spring* struck a nerve, for it was like no book written by male experts in the field. It had a plot as well as characters; Carson's birds and trees were anthropomorphized to tell their story, and the spring spoke volumes through its silence. Academics, chemical executives, the U.S. Department of Agriculture, and even the American Medical Association launched attacks against Carson, calling her book an unpatriotic polemic in the light of American competition with the Soviets for technological supremacy. Agents of Cold War politics tagged her with the gender-loaded term "hysteric," as if they weren't creating hysteria of their own. William Darby of the Vanderbilt School of Medicine called her environmental warnings "high pitched sequences of anxieties," not unlike the rants of housewives. Defending Carson, one scholar thought that "Francis Bacon would have been proud of such a manifesto advocating man's role as conqueror, master, and controller of nature. . . . Darby was speaking as someone whose power was being undermined."[14]

Carson was shy and reclusive, but after writing the book she courted the attention of media outlets, government officials, the average person on the street. When sales of *Silent Spring* reached the half-million mark, other investigators sought to confirm her claims about DDT for themselves. *CBS Reports* aired an hour-long special, and President John F. Kennedy appointed a panel to explore pesticide use. Before long, grassroots organizations were lobbying for regulation of the industry. When Carson testified before Congress, Senator Abraham Ribicoff welcomed her much in the same way Lincoln did Harriet Beecher Stowe when she came to the White House in 1862. "Miss Carson, you are the lady who started all this," Ribicoff told her; he didn't refer to a war between states, but rather to an ideological battle between big science and allies of the environment. Kennedy's Science Advisory Committee vindicated Carson before she succumbed to breast cancer in 1964.[15]

Carson's critics were widespread, for many felt threatened by her gender, her ambiguously scientific persona, and the resonance of her message. Her specific warnings about pesticides may have been less potent than the underlying statement she made about American progress and male dominion by extension. The tone of *Silent Spring*—sometimes poetic, never didactic, and always accessible—helped to move popular consensus away from the veneration of masculine scientists, who seemed to dominate and control. Few would have predicted that twenty years into the postwar period a heterogeneous group of pacifists, ethicists, environmentalists, feminists, New Leftists, and relativists of all stripes would question the very foundations upon which scientific progress had been measured. Could we applaud technological advancement that came at the expense of social equality, clean air, moral righteousness, or human rights? From militarism to imperialism, car emissions to napalm, the ethics of technical proliferation began to enter the minds of average Americans, leading to their growing mistrust of institutional science. The scientist as idealized in 1953 was timeless and virile. A decade later many viewed him as a dehumanized cog in a machine in which womanly sociability could and should be valued over false ideals of virile objectivity.

New conceptions of science did not immediately transform the experience of women in scientific institutions. The effects of the 1940s and 1950s were still apparent in the few women training or securing jobs in science fields in the 1960s, and the relatively large numbers of women in science during the 1920s were not matched until the 1970s. At the turn of the twentieth century Ellen Swallow Richards and other women went into home economics to be able to work as scientists. Lillian Gilbreth made similar choices in the 1920s and 1930s, and women were still following the same paths in 1960. At Cornell University, where more women were on the faculty than at twenty other leading universities, all but four of its eighty-seven female faculty members held appointments in its home economics department. At MIT and Princeton, where there was no such department in 1960, there were also no women faculty at the assistant level or higher. When Margaret Rossiter looked for research associates and lecturers teaching courses at these and other elite institutions, however, she found women in abundance—scientists who, incidentally, went on to win Guggenheims and to become presidents of their respective professional societies. Territorial and hierarchical segregation was alive and well.[16]

When Curie came to the United States, one out of every twelve American doctorate degrees in math and physical science and one out of every six in biological science were awarded to women. During the 1940s and 1950s the ratios of women to men in math and physical science plummeted to one in twenty-five, and in life science one in eleven.

In the 1960s actual numbers of women in science fields rose, but their proportion to men decreased or barely changed. In engineering, there was approximately 1 woman for every 114 men in the field. Only in biological science did women maintain a significant presence (27 percent), but even here, as in all science fields, the presence of women diminished as one viewed the rising promotional ladder. One could count on one hand the number of women among the more than seven hundred members of the National Academy of Science. In the 1960s men found jobs more readily than did women, and they earned $1.50 to every dollar earned by female colleagues. While four out of five male scientists were married and living with their spouses, only two out of five women were in a similar situation. Marital status was not an obstacle for men at any level, but the proportion of married women plummeted among those holding more than a bachelor degree.[17]

The inertia in science hadn't changed, but 1962–63 proved to be as transformative a moment in feminist politics as it was in the conception of scientific epistemology. Maria Mayer received her Nobel Prize in the virile field of atomic physics at the same time that Betty Friedan demythologized the sanctity of women's suburban lives in *The Feminine Mystique*. Friedan's account of the frustrations of middle-class housewives moved rapidly into public consciousness—as swiftly as Carson's warnings about DDT. President Kennedy's Commission on the Status of Women confirmed the cultural paradox Friedan had uncovered: at no other time in the nation's history had women been so committed to the full-time rearing of children and to their own professional endeavors. Tacitly, the commission acknowledged that American women were working women, often out of necessity, and that child care was a primary cause of inequality in the workplace, since children, as well as all things domestic, were assumed to be the singular burden of women, whether or not women worked outside the home. No one really challenged this premise, but the findings were integral to Congress's passing the Civil Rights Act of 1964, which made it illegal to discriminate on the grounds of gender in any job sector.

Equality between male and female scientists had never been a social reality, yet more women scientists began to think it a goal worth pursuing. Women at MIT asked students and faculty around the country to join them for a symposium, "American Women in Science and Engineering," and they received a massive response. In 1964, 260 student delegates from 140 colleges and 600 other guests, including professional women scientists, college deans, and high school guidance counselors, convened in Cambridge to hear Chien-Shiung Wu and eighty-six-year-old Lillian Gilbreth talk about parity in science. Panelists were critical of the culture of science, but not so critical as to challenge its androcentric foundations. Bruno Bettelheim of the University of Chicago didn't question expectations of women as mothers and homemakers; rather, he argued that women brought a valuable eye to scientific inquiry. When he described women as masters of "inner space," and men, of "outer space," he meant literally that men dominated the study of stars as well as all that is external—"nature at large." Women could be effective scientists, he determined, but their intellectual drive didn't surpass their deeper needs to be mothers and companions to men. Sociologist Jessie Bernard, too, cited biological proclivities for women's alleged attraction to teaching as opposed to research. Women are nesters, she implied; chairs of engineering departments could rest assured that female faculty would tend to students and "mind the store," while male faculty ran experiments in the lab.[18]

The MIT conference made clear that women were serious about confronting sexism in science, even if their views of problems were often essentialist and one-sided. Shortly afterward Betty Friedan and others organized the National Organization for Women (NOW) to see that the Civil Rights Act was upheld in the workplace, laboratories included. Seizing on the impetus to organize, women scientists formed caucuses within their professional organizations. The Committee on the Status of Microbiologists (1970), Women in Science and Engineering (1971), the Women's Committee of the American Physical Society (1971), the Women's Caucus of the American Association for the Advancement of Science (1971), Women in Cell Biology (1971), the Caucus of Women Biophysicists (1972), the Committee on Women in Biochemistry (1972), the Task Force on Women in Physiology (1973), the Women Geoscientists Committee (1973), and the Women's Caucus of the Endocrine Society (1976) were all spawned to police the hiring and promotional practices of institutions. Their preoccupation with breaking glass ceilings and getting more women on promotional ladders characterized the liberal agenda of early second-wave feminism.[19]

The caucusing of women scientists coincided with the passage of Title IX of the Education Amendments and the Equal Opportunity

Employment Act of 1972. The national thrust toward affirmative action brought increased attention to female recruitment in the National
Science Foundation, several prominent appointments of women scientists in Washington, and the passage of Senator Edward Kennedy's
bill to promote women in science in 1978. Citations of medical, psychological, and sociological reports on institutional discrimination, "math
avoidance," and "math anxiety" turned commonplace on the pages of
Ms., *Time*, and the *Wall Street Journal*. At the same time, female scientists—from Marie Curie to Jane Goodall—were being transformed into
literary heroines and role models for adolescent girls thinking about
careers as scientists.

Popular attention to the "woman-in-science question" lured women
to science fields, to be sure. Over the decade of the 1970s the number of
female science doctorates rose from 9 to 21 percent of all degrees conferred. Still, women's job prospects, salaries, and promotions did not
equal men's. The rate of unemployment remained four times higher
for female scientists, and though the biological, social, and behavioral
fields embraced women, even in these the paths of men and women
diverged with more time out of school. In 1979 the National Academy
of Science reported that the largest group of women scientists were assistant professors (31–38 percent), while the largest group of their male
counterparts were full professors (37–45 percent). Even on the tenure
track, female faculty were more likely to be asked to head lunchroom-
improvement committees than committees devoted to funding and
publication. In the fifty most prestigious science programs, 30–40 percent of women occupied positions completely off the promotion ladder. Between 1982 and 1989 more than 20 percent of women scientists
and engineers left their jobs. As late as 1993, women holding science
doctorate degrees earned 20 percent less than men, and the 2007 report *Beyond Bias and Barriers* confirmed that parity had still not been
achieved by any empirical measure. In 2008 the membership of the
National Academy of Science was still disproportionately male: 2,464
men to 232 women. "The glass ceiling has been raised," wrote Vivian
Gornick in 2009, "but it is still in place."[20]

Feminists had raised awareness and fought to pass legislation,
so what was keeping the woman scientist down? MIT physicist Vera
Kistiakowsky, the first chair of the Women's Committee of the American
Physical Society, determined that they had only scratched the surface

of the problem. She was reminded of the inaugural meeting of the Women's Committee in 1971, when a prominent male physicist brashly insisted that he, too, would have been a Marie Curie had he been married to someone like her husband. Laws and affirmative action were fine as far as they went, Kistiakowsky explained, but they weren't going to alter this man's ingrained belief that women hadn't the scientific talents of men.[21]

Kistiakowsky touched on the limitations of some feminist views. In aiming to right the absence of women in American science, such analyses never challenged the bias of science or that its model of professional success was predicated on a sexist notion of separate spheres. Biographers of important women scientists imagined them in masculine molds, the success of a token few defined as finding a way into the rarified category of "Great Men of Science." Even feminist biography tended to perpetuate the Curie complex, which had plagued women since Curie had first stepped onto U.S. soil. Women scientists were supposed to work harder and longer to compete with men, and women understood also that they remained the primary housekeepers and caregivers after grueling days in the lab. Unless women were superwomen, how were they supposed to do it all? The most successful men in science—Einstein, Oppenheimer, and most others—had relied on women's unpaid labor in the home to free them for thinking intensely about scientific problems. Even successful scientific women—Curie, Mayer, and sadly only a few others who come to mind—relied on paid domestic workers to give them the luxury to think.

Some feminists expected women to be female at home and male in the lab. Other feminists doubted that turning women into men was desirable, for they didn't think idealized male scientists were worth emulating. Genevieve Lloyd was dubious about "the maleness of the Man of Reason." Similarly, Shulamith Firestone thought that "the public image of the white-coated Dr. Jekyll with no feelings" was dangerous: "by separating his professional from his personal self, by compartmentalizing his emotion," such a scientist was left unnaturally divided between his private and public selves. He could not "integrate the amazing material of modern science with his daily life," she determined; he attended church as he built bombs and saw no contradiction. Firestone's catalog of scientific vices (the atomic bomb was prominent on the list, as was the medical manipulation of women's bodies) duplicated the catalog of "male" vices in American life. "This is to be expected," she explained. "If the Technological Mode develops from the male principle then it follows that its practitioner would develop the warpings of the male personality in the extreme."[22]

Few women scientists looked at their male peers or Western science in so sinister a light, but more of them began to critique the literal manmadeness of science. Their awakening occurred in American culture generally, as they convened to discuss their oppression in the bedroom and the workplace, the kitchen and the laboratory, and discovered that their complaints were more than individual hang-ups, but were collective crosses to bear. Women developed a consciousness as an oppressed class, and they saw that the radical mantra, "the personal is political," operated in many contexts. The private and the public, domestic and professional, emotional and objective, feminine and masculine, were not in fact inherently opposed but rather imagined that way to create hierarchies of power. In seeking to appear perfect in both professional and domestic worlds, feminist careerists were complicit with the fiction that professionals—scientists included— played no role in producing babies, rearing children, or fulfilling other social responsibilities.[23]

At the Conference on Women in Science, Social Science, and Engineering at Purdue University in 1981, panelists still sought strategies for placing women in male institutions, but they were also critical of those who ignored the departure of women from the science pipeline after being socialized like men. They viewed gender inequality as a theoretical and systemic phenomenon, the "woman-in-science problem" as culturally contingent. Sandra Harding later articulated this more radical position in encompassing terms:

> [It] holds that the epistemologies, metaphysics, ethics, and politics of the dominant forms of science are androcentric and mutually supportive; that despite the deeply ingrained Western cultural belief in science's intrinsic progressiveness, science today serves primarily regressive social tendencies; and that the social structure of science, many of its applications and technologies, its modes of defining research problems and designing experiments, its ways of constructing and conferring meanings are not only sexist but also racist, classist, and culturally coercive.[24]

Feminist scholars argued that women's alleged lack of scientific ability was not proven but imagined; nurtured, not natural. That didn't mean, however, that differences between male and female proclivities and orientations didn't exist. If cultural definitions of masculinity had functioned to keep the woman scientist down, would a reassertion of culturally female traits—relational and contextual thinking, and co-operative and intuitive styles—rather than hierarchical and unidirectional orientations, lend stature to women scientists and fresh perspective to scientific practice? A large group of social and physical scientists thought yes, and described women's ways of knowing as caring, holistic, and maternal without apology. They challenged their colleagues to find value in the knowledge gained in women's traditional activities and spheres of influence. What medicinal insights, for example, could be found by taking seriously the knowledge gained through generations of mothers teaching daughters herbal remedies in the sickroom? What were the benefits of revaluing the "hands-on" science of women in non-professional settings? These scholars, grouped together as "difference feminists," maintained that women's different cultural conditioning caused them to ask different scientific questions in a different scientific voice for the good of science itself.[25]

Many feminists then and since have warned of the dangers of over-determining gender differences, for stereotypes of "universal woman" can be as troubling as time-honored notions of "universal man." Sociobiology, a biological rationale for gender difference, was especially polarizing when it grew in popularity in the 1970s and 1980s. In the 1990s the question of creating a "feminist science" was rephrased: "Do women do science differently?" And in 2004 new debates ensued in *Science* over whether or not there was a "female style" in the lab. That Larry Summers' comments elicited such widespread response in 2005 indicates, if nothing else, that the nurture/nature debate in science is far from closed.[26]

Notes

1. Naomi Weisstein, "'How Can a Little Girl Like You Teach a Great Big Class of Men?' the Chairman Said, and Other Adventures of a Woman in Science," in *Working It Out: 23 Women Writers, Artists, Scientists, and Scholars Talk About Their Lives and Work,* ed. Sara Ruddick and Pamela Davis (New York: Pantheon Books, 1977), 242–50.

2. Bernice T. Eiduson, *Scientists: Their Psychological World* (New York: Basic Books, 1962); Paul M. Siegel and Peter Rossi, "Occupational Prestige in the U.S., 1925–63," *American Journal of Sociology* 70 (November 1964): 286–302.

3. Eiduson, *Scientists*, 151–52, 172–73; Walter Hirsch, *Scientists in American Society* (New York: Random House, 1968), 22.

4. Harriet Zuckerman, *Scientific Elite: Nobel Laureates in the United States* (New York: The Free Press, 1977), 176; Burton Feldman, *The Nobel Prize: A History of Genius, Controversy, and Prestige* (New York: Arcade Publishing, 2000), 18.

5. Hirsch, *Scientists in American Society*, 34–36; "The Image of the Scientist in Science Fiction: A Content Analysis," *American Journal of Sociology* 63, no. 5 (1958): 512.

6. Hirsch, *Scientists in American Society*, 92. There are some discrepancies in the numbers, though all support the claim that expenditures rose precipitously in these years. Alice Rossi claimed that the national expenditure on scientific research was $14.7 billion in 1961–62 and $16.4 billion in 1962–63. See "Barriers to the Career Choice of Engineering, Medicine, or Science Among American Women," in *Women and the Scientific Professions: The MIT Symposium on American Women in Science and Engineering*, ed. Jacquelyn A. Mattfeld and Carol G. Van Aken (Cambridge, MA: MIT Press, 1965), 56; Robert Kargon, *The Maturing of American Science* (Washington, DC: American Association for the Advancement of Science, 1974), 205.

7. Ralph Lapp, *The New Priesthood: The Scientific Elite and the Uses of Power* (New York: Harper and Row, 1965), 2; Mary Palevsky, *Atomic Fragments: A Daughter's Questions* (Berkeley: University of California Press, 2000), 1–4.

8. J. Robert Oppenheimer, "Physics in the Contemporary World" (lecture given at MIT on November 25, 1947), in *The Open Mind* (New York: Simon and Schuster, 1955), 88; Helge Kragh, *Quantum Generations: A History of Physics in the Twentieth Century* (Princeton, NJ: Princeton University Press, 1999), 394–408.

9. Catherine Rampell, "The Atom Spy that Got Away," *NBC News*, August 13, 2004, http://www.msnbc.msn.com/id/5653644; "Facing Life," *Time*, August 9, 1954, http://www.time.com/time/magazine/article/0,9171,936255-2,00.html; Alice Kimball Smith, *A Peril and a Hope: The Scientists' Movement in America, 1945–1947* (Cambridge, MA: MIT Press, 1970), vii (originally published by University of Chicago Press, 1965.)

10. Thomas Kuhn, *The Structure of Scientific Revolutions* (Chicago: University of Chicago Press, 1962).

11. Some of the sociological and philosophical studies spawned by Kuhn's work include William Kornhauser, *Scientists in Industry* (Berkeley: University of California Press, 1962); Derek Price, *Big Science, Little Science* (New York: Columbia University Press, 1963); Karl Hill, *Management of Scientists* (Boston: Beacon Press, 1964); Don Price, *Scientific Estate* (Cambridge, MA: Belknap Press of Harvard University Press, 1965); Warren Hagstrom, *Scientific Community* (New York: Basic Books, 1965). See Herbert Marcuse, *One-Dimensional Man: Studies in the Ideology of Advanced Industrial Society* (Boston: Beacon Press, 1964), xli; Paul Feyerabend, *Against Method* (New York: Verso, 1993), 7–8 (originally published with New Left Books, 1975); Kragh, *Quantum Generations*, 402–4.

12. Ruth Ginzberg, "Uncovering Gynocentric Science," in *Feminism and Science*, ed. Nancy Tuana (Bloomington: Indiana University Press, 1989), 81–82.

13. Rachel Carson, *Silent Spring* (Boston: Houghton Mifflin, 1994), xviii–xix; Frank Graham Jr., *Since Silent Spring* (Boston: Houghton Mifflin Company, 1970), 15–47.

14. Michael B. Smith, "'Silence, Miss Carson!': Science, Gender, and the Reception of *Silent Spring*," *Feminist Studies* 27, no. 3 (2001): 733–52; William Darby, "Silence, Miss Carson!" *Chemical and Engineering News* 40 (October 1, 1962), 60–62; Graham, *Since Silent Spring*, 48–65.

15. See Al Gore, introduction to Carson, *Silent Spring*, xix, xv–xxvi; Patricia Hynes, *The Recurring Silent Spring* (New York: Pergamon Press, 1989), 16–20; Graham, *Since Silent Spring*, 74–79.

16. Margaret Rossiter, *Women Scientists in America: Before Affirmative Action, 1940–1972* (Baltimore: Johns Hopkins University Press, 1995), 128–33, 155.

17. Rossi, "Barriers to Career Choices," 58–60, 66, 73–74.

18. Lillian Gilbreth, "Closing the Gap," 217–31; Bruno Bettelheim, "The Commitment Required of a Woman Entering a Scientific Profession in Present-Day American Society," 3–19; Jessie Bernard, "The Present Situation in the Academic World of Women Trained in Engineering," 166, 163–82, in *Women and the Scientific Professions*; Lilli Hornig, "Professional Women in Transition," in *Women in the Scientific and Engineering Professions*, ed. Violet B. Haas and Carolyn C. Perrucci (Ann Arbor: University of Michigan Press, 1984), 44–46.

19. Anne M. Briscoe, "Phenomenon of the Seventies: The Women's Caucuses," *Signs: Journal of Women in Culture and Society* 4, no. 1 (1978): 152–58.

20. Michele L. Aldrich, "Women in Science," *Signs* 4, no. 1 (1978): 126–35; Betty M. Vetter, "Changing Patterns of Recruitment and Employment," in *Women in the Scientific and Engineering Professions*, 61; "Women in the Natural Sciences," *Signs* 1, no. 3 (1976): 718; Hornig, "Professional Women," 53; Londa Schiebinger, *Has Feminism Changed Science?* (Cambridge, MA: Harvard University Press, 1999), 35, 61; Vivian Gornick, *Women in Science: Then and Now* (New York: The Feminist Press, 2009), 103–4.

21. Vera Kistiakowsky, "Women in Physics: Unnecessary, Injurious, and out of Place?" *Physics Today* 33, no 2 (1980): 32.

22. Sandra Harding, *The Science Question in Feminism* (Ithaca, NY: Cornell University Press, 1986), 79–80; Genevieve Lloyd, *The Man of Reason: Male and Female in Western Philosophy* (New York: Routledge, 1993) (originally published in 1984); Shulamith Firestone, *The Dialectic of Sex: The Case for Feminist Revolution* (New York: Bantam, 1970), 182–83.

23. Schiebinger, *Has Feminism Changed Science?* 92.

24. Harding, *The Science Question in Feminism*, preface, 224.

25. Ruth Hubbard, "Science, Facts, and Feminism," in Tuana, *Feminism and Science*, 119–31.

26. Londa Schiebinger, "Has Feminism Changed Science?" *Signs*, 25, no. 4 (2000): 1171; *Has Feminism Changed Science?*, 8; Ruth Bleier, *Science and Gender: A Critique of Biology and Its Theories on Women* (New York: Pergamon Press, 1984), 2–5, 194–95.

6 Generational Divides:
Rosalyn Sussman Yalow, Evelyn Fox Keller, Barbara McClintock, and Feminism after 1963

> In a science constructed around the naming of object (nature) as female and the parallel naming of subject (mind) as male, any scientist who happens to be a woman is confronted with an a priori contradiction in terms. This poses a critical problem of identity: any scientist who is not a man walks a path bounded on one side by inauthenticity and on the other by subversion. Just as surely as inauthenticity is the cost a woman suffers by joining men in misogynist jokes, so it is, equally, the cost suffered by a woman who identifies with an image of the scientist modeled on the patriarchal husband. Only if she undergoes a radical disidentification from self can she share masculine pleasure in mastering a nature cast in the image of woman as passive, inert, and blind.
>
> —Evelyn Fox Keller, *Reflections on Gender and Science*, 1985[1]

THERE WAS NO GREATER COMPLIMENT THAN THE ONE A JOURNALIST paid Rosalyn Sussman Yalow when she dubbed her "a Madam Curie from the Bronx" in 1978. After she became famous, Yalow hosted a Public Broadcasting Service series on the French scientist, for one of her greatest inspirations had been a biography written by Curie's daughter that she read as she embarked on her academic career in 1937. "For me," she reflected, "the most important part of the book was that, in spite of early rejection, she succeeded. It was in common with my background, with my being aggressive." Yalow also would see herself as a mother, a physicist, and a come-from-behind figure, one who, as a woman, had to work harder and longer to come out ahead. In graduate school she was inspired yet again by the film adaptation of Eve Curie's book: "Till my dying day, I will remember Greer Garson and Walter Pidgeon . . . coming back to the laboratory and seeing the glowing that meant they had discovered radioactivity." Yalow's later discovery of radioimmunoassay (RIA) played out in her mind in the same romantic way; depictions of Curie's life and work had been her template as well as her lens for viewing herself.[2]

Rosalyn Sussman Yalow, too, was the poor girl who made good, and had a Nobel Prize to prove it. "No one could have deflected me from my path," she later told the Nobel Foundation. Like Curie, she didn't sulk over hardship but tried harder, skipping grades until she proved she deserved scholarships to study science. She learned to read before kindergarten. When math anxiety set in for other girls, she had already decided to pursue calculus and chemistry. She came from a working-class family and contributed to the family wage by cutting out patterns for her uncle's necktie business. When there were no books in the house, she borrowed from the public library; when technical classes weren't offered at school, she sought them out elsewhere. Her parents did not have a high school education, but she never considered not going to college. Decades later she transformed her personal story into a parable about becoming the agent of one's destiny through hard work, not excuses or handouts. A sign hung prominently in her office that read: "Whatever women do, they must do twice as well as men to be thought half as good." This wasn't a complaint, but an assertion; she excelled because she had always been up to the challenge.[3]

Like other gifted daughters of New York immigrants, Rosalyn Sussman made grades that were good enough for her to enter Hunter College, where in those days tuition was free. She slipped into a lecture hall to hear guest Enrico Fermi speak about radioisotopes in 1939 and felt awed. To an ambitious girl from the Bronx, physics seemed the route to excitement and prestige, and thanks to her dogged persistence, administrators created a physics major. Sussman was the first to complete the program in 1941, but faculty members made no calls to graduate schools on her behalf. Acting on advice, she took stenography courses and a secretarial job at Columbia University's College of Physicians and Surgeons, hoping to gain access to the college labs. Then she accepted a teaching assistantship in the physics department at the University of Illinois, probably once reserved for a man who had gone to war. There were no women's restrooms in the lab buildings, and Jews were not allowed to live in on-campus housing. When Sussman appeared on campus in September 1941, she was a homeless woman among four hundred men, but she was completely undeterred.[4]

At the Sorbonne Marie Curie had met Pierre. At Illinois Rosalyn Sussman met Aaron Yalow, the son of a rabbi and also a graduate student entering the physics group. Like Pierre, Aaron didn't have the competi-

tive spirit of his wife, and friends and colleagues viewed him as emasculated in her hands. The Yalows' son believed that his dad stepped aside
to let his mom's career blossom because he worshipped the ground that
she walked on, but Rosalyn claimed that Aaron's Orthodox Judaism
was to blame for his less illustrious career. She had no qualms about
working on the Sabbath and High Holidays if that was what it took to
get ahead. Nothing was going to derail her. At Illinois she persevered,
seemingly impervious to criticism. When she got an A- in her optics lab,
the physics chair decided that the minus was proof that women were
no good at bench work. Yalow ignored him and moved on. She had a
heavier teaching load than her peers and after 1943 took on the added
burdens of keeping house and a kosher kitchen at the peak of wartime
rationing. She and Aaron had finally married, delayed only by university nepotism policies.

Yalow had no intention of letting marriage interfere with her plans.
She collected her PhD in nuclear physics in 1945 and took a job as an assistant engineer at the Federal Telecommunications Laboratory. Despite
the shortage of technically trained men, she was the only woman engineer at the company, and she was replaced when qualified men returned
from war. She returned to Hunter to teach physics, moonlighted in labs
without pay to keep her bench skills sharp, and never doubted that her
state of limbo was temporary: "I, as a small child, made up my mind
that people and institutions were going to need me, and I was going
to let them know." One can already detect the swagger some admired
and others detested when she won the Nobel Prize thirty years later.
Her determination was essential in a field as grim as postwar physics
for women.[5]

Aaron introduced his wife to Edith Quimby, a pioneer in the medical use of radioisotopes; that Quimby was also a mother of four cast her
further into the Curie mold. Yalow thrived in Quimby's lab, if only as a
volunteer. She was put in touch with the director of radiotherapy at the
Bronx Veterans Administration Hospital, who offered her a part-time
consulting position in a new radioisotope service. Yalow showed her
moxie when she took over the janitorial closet and built radiation-de-
tection instruments from scratch. By 1950 she had dropped all teaching
to work at the VA hospital full time.[6]

Yalow's dream of rearing children had been postponed but not ex-
tinguished. If her goal seems incompatible with her professional ambi-

tions, one may see it as a response to the "old maid" professors at Hunter who expected bookish girls to become schoolmarms and low-grade spinster scientists. Yalow wanted to be an exception, even as she, like most educated women, felt postwar pressures to keep the home fires burning. She couldn't ignore the opportunities that wartime technologies had opened up for scientists with her skills and ambitions, and thus she vowed to do it all at once.

Like her French idol, she bore two children during a frenetic period in her career—Benjamin in 1952 and Elanna in 1954. The policy of the VA hospital called for women to resign once they reached their fifth month of pregnancy. The veterinarian in the animal lab left on these grounds, yet no one dared to approach Yalow. "I was too important," she figured. "I wasn't concerned because they needed me." She worked until the day before she delivered her son and returned a week later. When she tried to do the same with Elanna, doctors forced her to stay in the hospital for eight days. On the ninth, she gave a scientific paper in Washington.[7]

Her goal was to make the transition between home and work life so seamless that it didn't look like a goal at all. The Yalows reared their children less than a mile from her lab, making it easier for her to get to the office early in the morning and to come home to prepare the children's lunch. After tucking them into bed she often returned to the hospital for a late night, only to do the same the next day. Curie's anti-natural path didn't always account for eating, and Yalow's didn't allow for sleeping. As babies her children slept by day and fed in the wee hours of the night when she was home to nurse. Although her mother and domestic workers helped her, she cooked, cleaned, and typed school papers in years when her male colleagues assumed few responsibilities at home. A seventy-hour workweek was standard, eighty to a hundred hours never out of the question; and yet the children had their vacations, their birthday parties, and their three square meals. When she had to work weekends she brought them to the lab to play with the animals—and what kids don't love animals?[8]

Yalow called her experiment in balancing work and family a huge success, but her adult children didn't always agree. Elanna believed that "emotion was lacking" and that her mother was less self-sacrificing than self-absorbed, her affection conditional, when in evidence at all. She wasn't a mother who soothed hurt feelings; she was terrible when confronted. Assertiveness was necessary in Yalow's profession, but Elanna hated that, in her mother's case, it came with blinders to her other faults. In theory, having a Nobel-winning scientist for a mother creates incentive for a child to achieve. For Elanna, it confirmed only that she couldn't compete. She sympathized with Eve Curie, the "artistic" child, who felt like an outsider when the family talked physics at the dinner table. Vying for a high-powered mother's attention was futile, and so, rather than study science, Elanna studied children and wished to have her own. Maria Mayer's daughter, Marianne, reacted to her mother's science much the same way: she married a scientist but swore off working outside the home. "Mother missed a lot of fun when we were small," she later reflected. "I won't want a housekeeper to hear the excited confidences of my children when they come home from school." Ben Yalow had science in common with his mother, and yet he still found it hard to get her to take notice of him: "She was, for as long as I can remember, driven by her career. . . . She essentially had two families: there was the one she had married into and raised, the biological and marriage family, and there was the scientific family. . . In the day to day aspects it was clear that the scientific one was the one in which she spent the predominant fraction of her time."[9]

Young Irene Curie once admitted to resenting radium, Marie's "other baby." Yalow's kids also competed with a discovery their mother referred to with similar affection in her Nobel speech. The lab was Yalow's nursery and a home away from home. She took special pride in nurturing her postdocs until they left the nest and made names for themselves elsewhere. Former students recalled her acting like mom in the lab, running experiments and doing household chores simultaneously. Upon winning the Lasker in 1976, she whipped up potato salad and roasted a turkey for all the staff. "Anyone who can run a laboratory should have no trouble managing a house," she insisted. It was essential that science and domesticity look compatible—no, analogous—if she were going to be viewed as a success.[10]

We can only imagine what Aaron Yalow was feeling as his wife "kept house" in the lab. He took a teaching position at Cooper Union and assumed more responsibilities at home, even as his wife claimed the domestic work as hers. He also accepted that she spent more of her time with another man, Sol Berson, a charming Renaissance type, whom Yalow referred to as the smartest person she had ever known. He was a resident in internal medicine at the VA hospital in 1950, when she sought him out for his clinical experience. For most of the following twenty years, they worked together day and night, desks facing, closed off in an office they shared in Building C of the hospital. Colleagues described their collaboration as a conventional marriage: they grew so accustomed to each other's thoughts that they communicated wordlessly. Miriam Berson was cordial but couldn't help but feel threatened; Yalow did little to allay her anxieties, seeming to be guilty of emotional adultery, if nothing else.[11]

Marie and Pierre Curie have been passed down to posterity as an utterly complementary couple —he as the thinker and she as the doer, "the muscle" of the pair; only in the context of an intellectual relationship is the "muscle" a mark signifying lower status. Colleagues made similar observations about Berson and Yalow. She understood the physics, he the physiology. She was deliberate and logical, he prone to "flashes of insight." She did the bench work, he the public relations, cultivating contacts with professional societies and journal editors. Colleagues assumed that Berson was the "thinker" because he represented their work eloquently at meetings and in papers. Despite her background in theoretical physics, Yalow was perceived as the experimentalist and subordinate. Although she had initiated their working relationship, it soon became clear who wore the proverbial pants. At professional conferences she accepted the mantra that women are seen, not heard: Berson represented their work and fielded questions during the Q and A, and she grudgingly mingled with medical wives. "I was living in the medical world, and I didn't have a medical degree," she rationalized. But her acceptance of the helpmeet position didn't stop at conferences. She made his lunch, poured his coffee, and organized his travel itineraries. Like a personal secretary, she made sure his manuscripts were typed and mailed. At home, Yalow never conceded an argument, but in the lab, Berson was always right.[12]

In Berson's shadow, she may have appeared the less competent scientist, but no one denied that she was one-half of an extraordinarily productive team. Together they introduced radioactive tracers into the bloodstream and detected infinitesimal amounts of antigen (often hormone) in the blood by measuring its ability to displace the binding of radioactively labeled antigen to its antibody. The technique effectively allowed them to tag red blood cells, electrolytes, iodine, and albumin to study the distribution of globins, blood volume, serum proteins, bone and muscle metabolism, and thyroid function. By 1959 their RIA provided a tool with which to screen for diabetes and, soon, also hepatitis B and thyroid disease. It was fast, inexpensive, and ripe for any number of applications, since it measured hundreds of biological substances. When it came time to patent, Yalow followed Curie's lead, making RIA accessible rather than profitable. "Too much money is disruptive," she resolved. If altruism had been part of Curie's legend, then she, too, would develop science for science's sake.[13]

All went well until Berson died unexpectedly in 1972. Although he had moved on to become the head of medicine at Mount Sinai Hospital, Yalow still considered him her collaborator and better half. She had grown dependent on him in ways that left her compromised when he was gone. Many people wondered how she'd get on without his skills, connections, and ability to win over critics. Like her French idol after the death of Pierre, Yalow appeared the helpmeet rather than the intellectual spark of a research partnership. Most agreed that RIA deserved a Nobel Prize, but Berson's death confused matters: since a posthumous prize for Berson was impossible, would she, too, be out of the running? Yalow thought about going to medical school to get more credentials but chose simply to work harder and longer, often a hundred hours a week, to earn recognition on her merits alone. She assumed the editing and lecturing that she had once left to Berson. By 1976, some sixty papers had come out of her lab; the fruit of her labor was a Lasker Award and membership in the National Academy of Science.[14]

In 1977, when she received the early morning call announcing her Nobel selection, she had already been at work for several hours and had seen that champagne was on hand at the VA hospital. If privately she had doubted she would win, she knew she would have had fewer doubts had Berson been alive awaiting the news with her. At the Nobel festivities journalists had little to say about the outward appearance of the male laureates, all in standard black and white; yet they dedicated whole columns to Yalow's choice of a brocade vest over a blue chiffon dress. The king of Sweden escorted her to the royal banquet as Aaron Yalow followed her, arm in arm with the queen. Yalow had been selected to give the address to university students, but the usher inadvertently compromised her moment of grandeur by fetching the wrong "Dr. Yalow" to take the podium. Aaron gestured to his wife and the error seemed quickly forgotten, but the innocence of the mistake made the moment a poignant one. Had Yalow's selection for the Nobel Prize silenced doubt that she deserved it? Had it changed the mind of men and women, boys and girls, who unconsciously assumed that great minds would always be male?

This was no time for doubt; Yalow was sure that her selection to speak was meant to inspire. This was the same year that U.S. congresswoman Bella Abzug convened the National Women's Conference in Houston to discuss glass ceilings and the Equal Rights Amendment. One month latert, it seemed fitting for Yalow to invite three female students to the Nobel ceremony—one from Hunter College, one from her high school, and one from her junior high—and to speak to her audience about sexism in science. Indeed sexism was evident, she told them, but it was an obstacle to rise above, not a problem to reform. Afterward some feminists seized on the fact that she used the platform of Nobel to make a statement about women in science. Others saw clear limits to her message; she spoke of "easing the path" for women, but not changing the path or the scientific community to which it led.[15]

The following year Yalow turned down the *Ladies' Home Journal* Woman of the Year Award because it highlighted distinctions between men and women, when excellence, as she viewed it, was in no way related to sex. After winning her Nobel Yalow grew vocal in her disdain for women's awards and affirmative action policies, which she viewed as official admissions of women's inadequacy. In 1981 she weighed in on the side of a white man who sued a medical school for reverse discrimination. He should be angry, she insisted: what's noble about women and minorities winning competitive spots regardless of their credentials? She preferred to be judged alongside the best men in the room and believed that equal consideration regardless of sex was possible. As president of the Endocrine Society she lambasted younger women who caucused within the organization: "It bothers me," she told a junior colleague, "that there are now organizations for women in science, which means they think they have to be treated differently from the men. I don't approve." Younger women told her to wake up and smell the sexism, to which she replied, "Personally, I have not been terribly bothered by it. . . . If I wasn't going to do it one way, I'd manage to do it another way." The younger physicist Mildred Dresselhaus tried to understand that Yalow's attitude was born of lived experience, not antifeminist sentiment: "Ros was one of those visionaries who went around saying that for women to be in physics they had to be smarter than men. That shows she recognized barriers."[16]

Yalow's harsh demeanor may have been her best protection from failure, but it did not endear men or women to her. After she won her Nobel Prize colleagues thought her "arrogant," "belligerent," "awesomely full of herself." She wore a replica of her medal dangling from her neck, a constant reminder that she thought herself better than others. She had reached the pinnacle of scientific accomplishment and boasted that she had done so with "help from men, not women." If she could commiserate with less fortunate women scientists, she wouldn't admit it, for she saw herself as a loner with whom no one could compete or compare.[17] Earlier in her career when Berson was around, she had mentored women and mothered young associates, but on her own she was known to be solitary. Her intensity was of a kind that Anne Roe once heralded as the hallmark of the heroic scientist, but a generation later this scientist seemed individualistic at the expense of fruitful collaboration. Of course one may also speculate about how much these negative views emerged from those envious of her reputation or disappointed about their own limited outcomes. As a woman, had she accepted her honors gratefully or gracefully enough? Curie had fallen prey to similar judgments.

Physicists in the Age of Women's Liberation

To younger women, Rosalyn Yalow was a contradiction: a model of masculine single-mindedness and also of glorified maternity; a scientific hero and, still, a woman who had tried to balance family and career. Hers is a story of rare triumph, but also inevitable sacrifice, even if she refused to acknowledge it. It's understandable that she thought her goals reachable only as a woman who transcended expectations of domestic women and scientific men at the same time. Ever since Curie's visit to the United States in 1920, female scientists had viewed the French legend as telling them to be more intense than men in the lab and unflappably feminine outside it. A variation on this image was that of the "superwoman" in the 1970s. She occupied all professional sectors, but she was particularly salient in science, thanks to sociological studies claiming that married women, if hardworking and organized, could be successful mothers and publishers of scientific research.[18] Yalow believed that women could have it all, and she scripted her own story as a tale of success achieved by working harder and being more focused—by taking Curie's anti-natural path.

If her contemporaries didn't choose similar paths, they often assumed her intensity and accepted her views about affirmative action and masculine science. Joan Freeman, a nuclear physicist who trained at Cambridge in the 1940s, tells a story much like Yalow's of her rise in science: both of them struggled for access to the technical training that was the birthright of boys, but this made them stronger. They watched female colleagues with less skill and determination falter around them. Yalow saw pregnant lab techs get fired, but not her; Freeman saw women deny marriages so they wouldn't lose their jobs, but she didn't. They both recalled moments in their careers when they were hard pressed to find women's bathrooms near their labs, and they remembered when they earned a fraction of the salaries of their male peers. Still, they insisted that sexism was not a factor in their careers.[19] They had internalized a masculine conception of competence; to suggest that science should be anything other than the enterprise that they had internalized was to threaten their very being.

Maria Mayer was also characteristic of this generation. Although she had suffered hardship in the androcentric atmosphere of postwar science, she had also succeeded in it and ultimately found it legitimizing. Even as she became a role model for younger scientists who embraced change, Mayer felt ambivalent about the change she unwittingly helped to create. After she had won the Nobel, the Women in Science program of the American Institute of Physics asked her to speak to teenage girls and college women about science careers. The most important piece of advice she had for her audiences was not about math or shop class; she told them that the key was to marry the right man—preferably a scientist in his own right—who would understand how it felt to be impassioned by science and thus would be accommodating of a wife's career.[20] Her advice can be read in any number of ways: Was she saying that marriage and motherhood shouldn't stop women from pursuing science? Or was she urging future scientists not to let science compromise a traditional home life? She also seemed to be suggesting that a woman's success lay in the hands of a supportive man rather than in her own efforts, that she needed his connections and permission to proceed. As much as the image of the career woman increasingly appealed to young women, she understood its limits. In 1964 Alice Rossi confirmed the dichotomy in the minds of female college graduates about their academic futures: Four out of five women she surveyed admired women such as Mayer who received prestigious scientific awards, but few of them aspired to similar goals for themselves.[21] Mayer sensed their anxiety and refused to speak of her career as a choice made in sacrifice of other things. In reality, we know that her sacrifices were at times very painful.

Amid nascent debates about women's work and middle-class domesticity in the early 1960s, the wide coverage of Mayer in the print media suggests that Americans were ready to embrace women's accomplishments in science and in careers generally. This was also true when the press covered Curie in the 1920s, though not so twenty years later, when the popular treatment of women scientists seemed to suggest that women who "indulged" in labwork were selfishly shirking their maternal duties. When Gerty Cori won her Nobel Prize in 1947, she was compelled to shield herself from interviews of any kind.[22] Twenty years before and twenty years later, journalists heralded the accomplishments of Curie and Mayer, but also defended their womanliness. While male winners of the Nobel were called "Dr.," these women were referred to as "Mrs.," perhaps to assure audiences that their important work was done in the kitchen, not in the lab. A journalist covering Mayer's trip to Stockholm in December 1963 assured readers that, like any housewife, Mayer spent the days before and after the Nobel ceremony getting ready for the Christmas holidays. While male recipients tended to official business, Mayer disappeared "to do a little shopping."[23] A writer for *McCall's* also reassured readers that Mayer, an "intruder in masculine territory," was first and foremost a gentle woman:

> In Professor Mayer, a tiny, shy, touchingly devoted wife and mother, who speaks so softly she can barely be heard, science and femininity have achieved an astonishingly graceful union. Last winter, at fifty-seven, she received the highest honor that the man's world of atomic-age science can bestow. But on her spectacular night of professional triumph in Stockholm, when the glittering Nobel medal became hers, Maria saw everything through the starry eyes of a romantic woman: "It was a fairy tale," she says. "The King of Sweden gave me his arm after the ceremony. And my husband, Joe, looked enchanting in his white tie and tails—he had borrowed the trousers from our son." Now, months later, the magic of it still brings a special light to Maria's bright blue eyes.

Mayer's winning the Nobel had been transformed into a "Cinderella story in science," in which she was a domestic heroine—someone who gave dinner parties for which menus and crystal were "lovingly planned." Her children were her greatest joy . . . she loved growing flowers . . . she was mesmerized by the decorative accents of the royals' home . . . her favorite word was "elegant!" Though she "walk[ed] with the giants of science, she [wa]s also a sweet, diffident and an utterly feminine woman."[24]

Upon men's discoveries, we hear of pats on the back or shouts of "Eureka," the end results of masculine contest and exploration. The terms with which the press has conveyed female discovery have been no less stereotypical. Marie Curie, we were told, stood entranced over her glowing radium as a mother would stand over a sleeping child. In 1963 journalists seized on language that they claimed Mayer used to describe her shell theory to her teen-aged daughter: "Think of couples waltzing around a room," she allegedly told her. "They spin as individual couples as they also orbit the ballroom. Some couples spin in one direction, and some the other, just like the electrons that orbit a nucleus—and everybody who has danced a fast waltz knows that it's easier to dance in one direction than another."[25]

There was her theory in Disney-like packaging to add romance to the Cinderella story. The metaphor allowed the press to divorce Mayer from the masculine culture of science, to make her appear a different breed of scientist, if a scientist at all. Twenty years earlier, journalists covering Los Alamos told Americans that the physicists who developed the bomb had the power to kill; Mayer had been part of that fraternity, and yet in 1963 her atomic science was made over to appear unthreateningly dainty. She tacitly accepted the images but chastised a contributor to the *Christian Science Monitor* for claiming that she liked to bake: "I have never in my life baked a cake, and my husband doesn't like them anyway."[26] This writer had made similar claims in print about Lillian Gilbreth that were no less imaginary. Such creative license suggests how little cultural attitudes about science and women had changed by the early 1960s. Female scientists weren't admirable except when painted in glowing domestic shades.

Mayer's careerism was easier to defend once feminists could read her personal story of antinepotism and underemployment as a cautionary tale of sex discrimination for younger scientists. Women in the

American Physical Society and other science organizations sought her out, but Mayer was a reluctant martyr: "The fact that I worked as a volunteer professor of physics at the University of Chicago was not due to the fact that I am a woman, but due to the fact that I am a wife. My colleagues accepted me as one of theirs. I never felt any discrimination because I was a woman."[27]

Younger scientists might have thought her naive or in denial, but her dismissive tone was typical of her generation. Cecilia Payne-Gaposchkin, six years Mayer's senior, said the same when asked how being a woman had affected her scientific career. Oddly, she recanted her statements the following day and named a list of offenses against her, including the many times she had been passed over for tenure because she was female. The younger astronomer Vera Rubin thought she understood: "These indignities bothered [Gaposchkin] only at times. More likely, throughout her life she realized that she was taking steps never before open to a woman, and accepted them as part of the difficulties of being first." In 2000, Rita Levi-Montalcini, the Nobel prize-winning biologist and three years Mayer's junior, was likely of the same mind when she drew a rosy picture of her career, though her 1988 memoir leaves room for other interpretations. She admitted that, for a time, peers had virtually ignored her discovery of nerve growth factor, or at least that *she* had discovered it. Her winning of the Nobel Prize thirty years later served as proof that they eventually had come around, so she dwelled on this rather than her trials.[28]

Mayer's Nobel Prize and Gaposchkin's appointment as the first female full professor at Harvard also seemed to soften hard feelings. When younger feminists asked them to relive their professional stories, they tended to offer milder versions of them. Mayer didn't want her elongated career path scrutinized, but then reluctantly agreed to assist Joan Dash in gathering evidence of her employment history for a collection of feminist portraits called *A Life of One's Own* (1973). What may have appealed to Mayer about the project was that Dash intended also to highlight the supportive husband who had helped her succeed. With slight tweaking, Dash's rendition of Mayer's companionate marriage pleased Mayer, since it did justice to Joe. But she remained uncomfortable with Dash's account of her career as victimization vindicated. She had no ax to grind, Mayer reminded Dash, since her male colleagues had fought gallantly to shore up her success.[29]

We don't know whether Dash's corrections were acceptable in the end, for Mayer died before the book appeared. Other than a handful of celebratory juvenile biographies, there have been few accounts of her life since Dash's, suggesting that her wishes to blend into the walls of masculine science have been respected. Posthumously, feminists have made a more compelling martyr out of Rosalind Franklin, the popularly recognized victim of *The Double Helix*. They have argued that Watson's demonizing signals the sexist hostility she must have endured through her brief life. But she was also a victim of a form of cultural invisibility, since few people in the 1950s could imagine a young, attractive woman foregoing marriage for the lab. Watson couldn't either, so he represented her as caricature rather than in three dimensions.

Correcting the record, biographers have insisted that Franklin wasn't the dowdy bluestocking Watson made her out to be. Her town was not London, but Paris. She wore Chanel, Christian Dior, and lipstick and enjoyed nightlife and red wines. One author divulged her measurements: bust 34, waist 27, hips 38—physical proof of her possible sex appeal, while others emphasized her athletic body and independent attitude to make the case for her modern resonance. She's been described as cosmopolitan and grounded, decadent and virginal. Commenting on the British Broadcasting Corporation docudrama *Life Story* (1988), Francis Crick thought it no coincidence that Franklin's character was the only one portrayed with her head down doing bench work. On the television screen, she appeared as the Virgin Saint of Science because filmmakers liked that image. Certainly, she never was the Norma Rae of science, and yet she has appeared as a self-assured woman with politicized sensibilities, as if she were knowingly ahead of her time.[30]

Franklin's Birkbeck colleague Aaron Klug was amazed by the huge number of voices that emerged to right the wrongs of *The Double Helix*, but he wondered whether new caricatures would become equally constraining, for they, too, rarely delineated the living, breathing woman he knew. "I can't say she was like a man, not like that at all," he told a biographer, "but one didn't think of her being particularly like a woman; she wasn't shy, or self-effacing—but she wasn't blustering, either." When the New York Academy of Science considered awarding Franklin a posthumous prize for DNA, Klug expressed concern about the emphasis placed on her gender over her science: "If she is to be honoured, it should be not so much as a 'woman of science' but for her crucial contributions in sorting out the A and the B forms. . . . The fact is Rosalind was never an active feminist, but simply evoked or created respect in her own right as a person, and I think she might have found some of the present attitudes somewhat distasteful."[31] It was clear to Klug that Franklin's gender and science could not be culturally reconciled into a notion of personhood, just as Maria Mayer's couldn't in 1963 or Marie Curie's in 1920.

Those women who became scientists during the war years and lived to see the beginnings of women's liberation—Maria Mayer, Rita Levi-Montalcini, Cecilia Payne-Gaposchkin, and Rosalyn Yalow—stood at a cultural crossroad, where Rosalind Franklin would have stood as well had she lived. They felt tremendous pressure to appear also as perfect wives and mothers to counterbalance their achievements as scientists. Levi-Montalcini handled the tensions by choosing a life of virtual celibacy. The Nobel winner Barbara McClintock chose this path, too, but suffered the constant scrutiny of men who thought her eccentric and unwomanly. It's not an accident that it took more than thirty years for the peers of these single women to award them Nobel Prizes for work they did in the 1950s. Married contemporaries won their Nobels more quickly, but they also paid a price. Mayer handled the tensions by underplaying the struggle between home life and science, and Yalow denied struggle altogether, boasting unabashedly about the perfection she had achieved in both worlds. For all her superiority, it's likely she dissembled pain or, as her daughter contended, blotted it out with alcohol, as Maria Mayer was said to do.[32]

In her book of interviews called *Women and Science: Then and Now* (2009), Vivian Gornick confirmed that Yalow could have been nearly any woman scientist of this generation to whom she spoke. These were wives who succumbed to nepotism rules or "de-sexed bluestockings" who chose all-encompassing careers over marriage. "There was about them a uniform remoteness of manner and expression," Gornick noted. "The pose was haughty and aristocratic, the speech guarded, the personality masked, defensive, unknowing. Each of them, in effect, said to me: 'Problem? My dear, I *had* no problem. I know what I wanted to do and I simply did it.' Surely they needed hard exteriors to withstand the environmental hazards of postwar labs, but Gornick, a younger feminist, didn't see as heroic the disavowing of struggle: "They had been treated like den mothers and mascots by their colleagues," she said of this generation, rather than as real, multifaceted women.[33]

Scientists of the following generation inherited their attitudes to some extent but were young enough to be influenced by the feminist movement. Fay Ajzenberg-Selove provides several interesting points of comparison with Yalow, since she held similar views before becoming politicized in the late 1960s. She had wanted to attend MIT as badly as Yalow wanted to attend Columbia, but her sex made it impossible, and thus she also turned to a midwestern university. She, too, didn't challenge the notion that succeeding in physics meant embracing its male rites of passage—the shop classes, the years of intense intellectual competition—to weed out the weak from the worthy. "When I began work in physics, only one in forty American physicists was a woman," she proudly recalled, and that was what appealed.[34]

Like Yalow, Selove felt that her relationships with men accounted for her success in physics; Marie Curie was practically the only woman she ever cited as a positive force in her career. "I thought that scientific work would be appraised logically and unemotionally," she explained. "I wanted to avoid emotional outbursts which I related to the behavior of artists (and of Mother), and not of scientists." Both she and Yalow claimed that their interactions with men were sexually benign until they met the scientific husbands who supported their careers. As they made professional names for themselves working with men who weren't their spouses (Yalow with Berson and Selove with physicist Tom Lauritsen), their colleagues speculated about their marital infidelities. Few around them could imagine nonmarital partnerships as viable or productive.[35]

When Selove was engaged to be married, the forty-nine-year-old Maria Mayer pulled her aside and poured her a whiskey before ranting about the plight of the married woman physicist. It left an impression, but Selove still believed that the most competent women could make family coalesce with career. She was an untenured member in the physics department at the University of Pennsylvania, operating in a hierarchical system in which power relations were readily apparent. At Argonne, the VA hospital, and Cold Spring Harbor, the Nobel laureates Maria Mayer, Rosalyn Yalow, and Barbara McClintock buried themselves in research and insulated themselves from the tribulations of other women. But at Penn, Selove was in her office whenever women students needed consoling. One graduate student came to her when her advisor had refused to help her find a job. When Selove confronted the advisor, he confirmed that the student was one of the best he had ever had, but he wouldn't waste his time on her behalf. It was then that Selove's feminism, as she described it, was ignited like the flick of a switch.[36]

Gloria Lubkin, Selove's former graduate student and an editor of *Physics Today*, suggested that she talk with Brian Schwartz of the Forum on Physics and Society, who was organizing the first meeting of women in physics at the national American Physical Society (APS) meeting in 1971. Selove agreed that it was good timing: it was the year that Mina Rees became the first woman president of the American Association for the Advancement of Science. In previous years too, Selove had taken issue with the "ladies registration" at the annual APS meetings, since organizers assumed it was for spouses of physicists rather than for physicists themselves. Selove moderated the meeting, and more than six hundred people attended, leading physicist Vera Kistiakowsky to form the Women's Committee of APS shortly afterward.

Selove became the first female officer of APS, ironically just as she was suffering a sexist tenure review. Colleagues decided that she was not "sufficiently active in nuclear physics" and that she was too old for tenure consideration at forty-six. Being married to another department member was also an unspoken stigma. Although antinepotism policies had technically been overturned, the prevailing sentiment in the department was that spouses weren't promotable on their own merits. She filed a complaint with the Equal Employment Opportunity Commission, and became one of the first people to use the quantifiable measure of "citation counts" to make her case. The only faculty member in the department who could list more publications than she had won a Nobel Prize, making her case still more compelling. The Human Relations Commission decided unequivocally in 1973 that she had suffered discrimination, and thus she became the second female fully tenured faculty member in Penn's College of Arts and Sciences.[37] Selove's family history and career path resembled Yalow's, but her views of women's rights had been changed sharply by the women's movement. She chose to bring the law to bear on a situation that Yalow would have expected to work itself out in the allegedly Darwinian way things did.

A physicist who assisted Selove with the APS meeting was none other than Yalow's contemporary Chien-Shiung Wu. When Yalow first started at the VA hospital in the 1950s, Wu was nearby at Columbia and had moved into an apartment two blocks from her lab. Wu often left her son with a nanny or fending for himself; graduate students admitted their discomfort when he called the lab at night to tell his mom that he was hungry and wanted her to come home, although most of the time she was too absorbed in experiments to heed him. Wu's intensity earned her a reputation as the "Madame Curie of beta decay" as well as the more sinister "Dragon Lady" of physics. She worked six long days a week, took fifteen-minute lunch breaks, and was unforgiving when graduate students took personal days. This was a woman who sent her husband to take an anniversary cruise on the *Queen Elizabeth* without her, too wrapped up in experiments to be interrupted. A decade later at the MIT Symposium on Women in Science and Engineering she continued to view science as an intense endeavor, but she rallied with women to debunk the notion that intensity was a man's trait alone, citing Marie Curie, Lise Meitner, and Maria Mayer as proof in her field of physics.[38]

Wu became an active feminist who did not seek to change the culture of the lab so much as to make women a part of it. The MIT physicist Mildred Dresselhaus, once Yalow's student at Hunter, hoped, if not to change the culture outright, then to equip women with the tools to succeed and perhaps to make changes from within. She, too, had been part of the Symposium on Women in Science and Engineering in 1964 and the APS meeting of 1971. She was tenured in 1968, but she never forgot how male colleagues joked about her late arrivals to the lab during the years she had to wait for the babysitter to arrive. When she first taught at MIT there were few women in her solid-state physics course, and those who showed up remained silent in the back row of the hall. Rather than deny inequities, she organized forums for women to talk with older scientists about the pressures of the lab and provided assistance to those applying for grants and promotions. She has seen it as her responsibility to advance the cause of women in science. "Ros didn't move in that direction," she lamented, "So I did that. . . . It had to be done."[39]

Dresselhaus thought that if women achieved "critical mass," some 10–15 percent of graduate and faculty positions in science departments, they could begin changing institutions for the better. Others, including former physicist Elisabeth Fraenkel, doubted that fundamental reform was possible. Fraenkel wrote Vera Kistiakowsky of APS in 1971:

> I am NOT interested in women like Mina Rees, any more than I am in Ralph Bunche. This is tokenism, and is the worst kind of discrimination. I am interested in the dozens and dozens of women like myself

who idealistically went through school, hoping to be able to contribute, and who were sooner or later—usually sooner—slapped down.

. . . The women in my age group almost uniformly did poorly. Some—even with Ph.D's—ended in libraries. Many, especially in microscopy, started their own businesses. We almost did. Others felt they had to take what was dished out to them—but since I am a very poor masochist, I couldn't hack this at all.

. . . Today I am an active leader in the NY area in the liberation movement—every phase of it—and of course especially in Women's Liberation. I counsel women—hundreds of them—from all over the U.S. I help them be proud to be female, and not to let the male power structure suppress them. I also tell prospective women physicists, chemist[s], engineers NOT to waste their time and injure their dignity. Until things get a lot better, science is NO place for a woman in the U.S.[40]

Angry epiphanies like Fraenkel's came to women of her generation fast and furiously, as discoveries were said to have come to their male peers. "For the first time it hit me that my life had developed as it had because I was a woman, and I'd made women's choices, and ended up where women in science end up," one of Gornick's subjects divulged. It took Elga Wasserman, a chemist, twenty years to realize that her graduate advisor at Harvard had been too busy earning his Nobel Prize to take her seriously. "He had lost all interest in my career. . . . Nobody advised or encouraged my husband and me to embark on a joint job hunt, although it probably would have been futile anyhow." Her husband joined the Yale faculty and she agreed to be a research assistant in the microbiology lab. She left when her first child was born, and no one urged her to stay. She looked on admiringly at her colleague Mary Bunting, who came to the lab at night after her husband returned from the office to put their four kids to bed. His assuming of domestic chores seemed to make all the difference.[41]

Bunting became the president of Radcliffe College, and Wasserman also went on to higher education administration, at Yale, after twelve years of part-time work in industrial labs and community colleges. Meanwhile, many of their contemporaries seethed as they tried to persevere in university labs. The only tenured woman in an Ivy League chemistry department told Gornick that she hated that she had spent twenty-three years watching male colleagues pass her by. Another woman graduated first in her class and couldn't find a job; her husband, who graduated in the top third, received fourteen job offers, so she spent a career accepting fellowships attached to his grants. Science had seduced her and then left her defiled. A woman who described her graduate mentor as a womanizer meant this literally. Another woman hated that in choosing science, she had taken a "vow of lifetime celibacy," as yet another lamented having to hide a pregnancy under a lab coat and telling no one when her child was born. Gornick's subjects didn't question that they had to appear sexless, freakishly devout to research, until women's liberation made them newly aware. As much as they wanted to point fingers at men, they admitted that their mothers were guilty of abetting: "My mother said, 'Be a scientist,' but she didn't really mean it. What she really meant was, 'Get married and teach high school biology so you can be a help and a pride to your husband.'"[42]

In the case of several of Gornick's youngest subjects, divorces brought their powerless positions into high relief; they became feminists when they had to build careers and reputations from scratch. They had learned from women before them and saw that their frustrations were caused not by personal failings but by ceilings that doomed them from the start. They understood that mechanisms of subtle discrimination led to more of them paying their way through graduate school and being assigned teaching assistantships as men were assigned to research projects full time. More of them demanded visibility and believed that they had the right to fight for it or to move on. They were now their own keepers, and science didn't define them. "It was the range of temperamental difference among them that set the young women off from the old," Gornick concluded. "Instead of concealing variety they seemed to assert it. In fact, these women seemed to reveal their individual selves through science rather than disappear into science as the earlier generation of women so often did."[43]

The Transformations of Evelyn Fox Keller and Barbara McClintock

In 1957 Evelyn Fox was a twenty-one-year-old college senior who had a mind for theoretical physics. She knew this made her an oddity. Other smart girls excelled in English, a course in which she got Cs growing up; she, on the other hand, had preferred sci-fi to classic literature, especially George Gamow novels, in which boys fell asleep in physics lectures only to dream up brilliant theorems themselves. Her dream was to work with Richard Feynman, the subject of her senior thesis at Brandeis, and for a time she thought it possible. There had been rumors that he was taking a university chair at MIT, where she was being recruited, but he decided to stay at Caltech, which still hadn't opened its doors to women. Recruiters from Stanford and Harvard came to her with funding; indeed there seemed to be a lot of it for the taking. The National Science Foundation funneled more than ever down to university programs, physics departments in particular, to recruit the best and brightest to compete with the Soviets. In the end she decided to go to Harvard—undergraduate advisors had told her to settle for nothing less.

Looking back, she remembered feeling incredibly buoyant before arriving. "I fell in love with theoretical physics, and . . . with the image of myself striving and succeeding in an area where women had rarely ventured." "I believed not only in the possibility of clear and certain knowledge of the world, but also in the unique and privileged access to this knowledge provided by science, and scientists, as arbiters of the truth—physicists were, of course, the highest arbiters."[44]

In her idealized version of graduate school, she had seen herself ruminating on space and time as she imagined Einstein had in his younger years. But the stark reality was that "operationalism" reined supreme. "In place of wisdom, I was offered skills," she reflected. Rather than theory, instructors stressed hands-on technique. Her papers were returned with red slashes throughout: too fixated on the arithmetic, no one had bothered to notice her mastery of concepts. With each assignment male students grew bolder, and she more unsure. Only later did she understand that "in addition to the techniques of physics, they, [men], were also studying the techniques of arrogance." Her first year was an exercise in internalizing the cultural ascribing of intellectuality to maleness. In lecture halls and campus dorms she was consistently reminded that physics was hard and that she was wrong to think she understood what she thought she understood; her lack of fear proved her ignorance. No woman at Harvard had ever completed a graduate degree in theoretical physics; the prestige of the program relied on its producing only male scientists.

There were two other women among the one hundred students in the physics program, but neither was as serious as Fox about her studies. Fox later regretted that she had then believed that being taken seriously as a physicist meant not identifying with other women. Meanwhile she was scrutinized—for her wardrobe rather than her mind. She became the object of laughter when men detected her ambition. Oral exams were a nightmare: her main examiner didn't show up and cited over-sleeping as the reason, although the exam was at 2:00 p.m. She passed, but couldn't get a faculty member to answer his office door to assist in directing her thesis. She knew she was unwanted and grew lonely. She attended parties, only to be confronted with the same question of "why?"—Why stick around? Why so desirous of masculine goals? Once she had been in love with the idea of being a physicist. Now she wondered why she wanted this for herself. By the middle of her third year, she decided to give up physics. She had already stopped going to class and liked the response: professors were affable, even sympathetic.

In the summer of 1960 she found refuge in the laboratories of Cold Spring Harbor, where her brother, a biologist, had a room to spare. There she met Max Delbrück and became fascinated with the DNA code Crick and Watson had left ripe for deciphering. Her excitement for science rekindled, she returned to Harvard and worked in molecular biology, where colleagues were congenial and respected her mind. She completed her doctorate degree in 1963 and took a job at New York University teaching theoretical physics. She also married Joseph Keller, the mathematician who had supervised her, before having two children and charting new ground in mathematical biology. With colleague Lee Segel she invented equations for describing how single-celled amoebas changed into multicelled organisms. As Segel went around the country generating excitement for their findings, Keller followed her husband to Stanford and grew bored at home. She decided to collect data on a new set of organisms—ones that suffered fates not different from hers. She started "thinking about women in science."

Women often describe their years away from science rearing children as impeding their work or ambitions, but Keller turned the time into an opportunity to expand her mind in ways she couldn't when tied to the lab. She read deeply in literature and philosophy and remembered this period of "blossoming radicalism" as the most joyful in her life. In 1974 she returned to professional life with a series of lectures in mathematical biology, slipping into them her new mathematical models indicating the paucity of women in science. She had never before considered the extent to which science was bound to the idea of masculinity. "A lifelong training had labeled that question patently absurd," she admitted, but now she told a room full of men that this was the crux of the problem of women's status in science. She became one of the first feminists to use cultural and social indicators to reveal the workings of gender in the practices, institutions, and metaphorical language of science. "The fundamental conflict—between my sense of myself as a woman and my identity as a scientist—could only be resolved by transcending all stereotypical definitions of self and success," she realized. "This took a long time, a personal analysis, and the women's movement."[45]

Keller wanted to understand how socially constructed notions of gender defined science for better or worse. This required an exploration of science through new female eyes, but whose? In the late 1970s feminists wrote books about the occasional success story, typically a female physicist with tenure or a Nobel Prize. But these biographies often were no different from those written from a male perspective. "See," they asserted, "this woman can do what men can do on men's terms." When Keller began to search for a biographical subject, Yalow had just won the Nobel Prize, but Yalow, a woman who had never questioned the masculinity of her field or her prescribed womanly duties at home, would never do as Keller's subject. Keller preferred the enigmatic geneticist Barbara McClintock, a woman who quietly kept to herself in the cornfields of Cold Spring Harbor. When Keller was there in 1960, McClintock had been tucked away, doing science on her unorthodox terms.

No one denied McClintock's brilliance, but few biographers saw a modern-day heroine in her before Keller approached her in the late 1970s. She was not the superwoman type. The work-family balance of her contemporaries was moot in her world; there was no balance, only science—at least as her early myth had been spun. Other successful women scientists had succeeded because they obeyed institutional rules. McClintock hated rules and most institutions; she never had a boss that she liked. Yalow and Mayer had been undeniably female in appearance; McClintock, on the other hand, looked androgynous. She appeared to Keller in her seventies even as she had appeared in her twenties: her wardrobe was ambiguously plain and she wore no skirts or makeup and kept her hair short; if there were any curves on her one-hundred-pound frame, she had never accentuated them. She was single and never had children, nor had she shown pangs for them. People speculated that she was gay, but most decided she was asexual, married to the corn plants she had spent a lifetime knowing.

McClintock's corn and microscopes were hardly the stuff of titillating biography, but Keller found her fascinating, turning her story into a piece for the *New Yorker* and then into a full-scale biography. She was not simply hero or victim; Keller believed that both success and marginality characterized her career. Because the eccentricities that male scientists gloried in had served to stigmatize McClintock, her scientific innovations consequently had been overlooked. But Keller didn't see her story as a cautionary tale so much as an alternative one. "She had long ago rejected the role of 'lady scientist,'" Keller noted, "but when she had a title normally reserved for a gentlemen academic, she did not fit that role either." McClintock defied social categories of "woman" and "scientist" and in the process demystified science itself. By highlighting McClintock's relationship to nature, a relationship that Keller did not call "female" or "feminist" but that others eventually labeled as such, she hoped to show an approach to science that transcended gender and was good for science as a whole.[46]

Keller's heroine was three years older than Maria Mayer. She had studied cytology and genetics in the School of Agriculture at Cornell and had earned her PhD in 1927. During the Depression jobs were hard to come by, but this was a woman who, by the age of thirty, had revolutionized her field. As a graduate student she had identified the morphology of all ten chromosomes in corn; as a postdoctorate researcher she and graduate student Harriet Creighton demonstrated the first genetic crossover in *Zea* strains. T. H. Morgan at Caltech believed even then that the young scientist was "the best person in the world" in maize genetics, and yet men continued to overlook her for university positions. She traveled to colleagues' labs on grants from the National Research Council, and on Rockefeller and Guggenheim Fellowships, but she had no permanent position to fall back on.[47]

McClintock thought her gender was to blame but was not dejected about her nomadic status, for she hated the schedules, teaching, and committee work that came with university jobs. Nevertheless when Lewis Stadler found her an untenured position at the University of Missouri, she recognized the folly in turning it down. She stayed there until 1941, but was visibly depressed most of the time. Men in the department assumed that she was there to teach just as other women were, but she found the lectures a nuisance and complained about her insufficient research facilities. On a night that she locked herself out of her lab colleagues watched the trousered woman climb up to and through a second-story window, simply more proof of her inappropriate eccentricity. McClintock was convinced that she would be fired, and in fact there was cause, since she was known to shirk her teaching duties when they conflicted with harvesting her corn. She was contemplating a new career in meteorology when the Carnegie Institution intervened and offered her a temporary position in its genetics division at Cold Spring Harbor. She would have her own cornfields and live on-site. The salary was modest, but there was no formal teaching attached. Convinced that she'd have less freedom anywhere else, she accepted the job and became a permanent fixture there for the next fifty years.[48]

Although McClintock lived on the ground floor of a visiting women's dorm, she was known to sleep in her cornfields or on a cot in her office when absorbed in her work. She took walks on the grounds along Bungtown Road, collecting flower parts for chromosome preparations. Colleagues recalled occasionally accompanying her and listening to her explain the variable pigmentation patterns on Queen Anne's lace or the flight patterns in flocks of birds overhead. But most of the time she was physically, emotionally, and intellectually alone. McClintock told Keller that her preference for solitude "began in the cradle." As far back as she could remember she preferred to read or sit alone, "thinking about things." She never felt the need for strong personal attachments and couldn't understand marriage, for instance, since she "never went through the experience of requiring it." The admission might have given other biographers pause, but Keller thoughtfully connected her solitude and science. This was how she achieved intimacy; her closest relationships were with nature, and hence she achieved a richer, more holistic understanding of the natural world.[49]

Keller found it hard to encapsulate McClintock's science in a single term because of the encompassing and integrated perspectives she was able to achieve: "For her," Keller wrote, "the smallest details provided the keys to the larger whole. It was her conviction that the closer her focus, the greater her attention to individual detail, to the unique characteristics of a single plant, of a single kernel, of a single chromosome, the more she could learn about the general principles by which the maize plant as a whole was organized, the better her 'feeling for the organism.'" In this sense McClintock was like Lillian Gilbreth—a "systems thinker" who saw interrelatedness wherever she looked. Rather than presume that a master control had been hardwired into genomes of corn, for instance, she believed that the interactions between genomes and their cellular environments served as the mechanism of control. It was not her practice to dissect parts of organisms and study them one by one under fluorescent lights; she looked for interactions in native settings to see nature on its terms. She didn't "press nature with leading questions," Keller explained, she "dwell[ed] patiently in the variety and complexity of organisms." It wasn't enough to amass data about a kernel of corn; she came to know its temperament, its life story—a process more intimate and creative than science-as-usual.[50]

McClintock's communion with nature was so intense that she actually lost herself to it. "I'm not there!" she explained to Keller. She was no longer cognizant of the position she was supposed to assume lording over it; she was nature's equal because she was part of the system's whole. "I found that the more I worked with [chromosomes] the bigger and bigger [they] got, and when I was really working with them I wasn't outside, I was down there. I was part of the system. . . . I actually felt as if I were right down there and these were my friends." She collapsed the distance between subject and object, inquirer and inquired, entering conversations with nature and waiting patiently to hear its utterances. Critics might call it unobjective science, but Keller saw it as science with context—"dynamic objectivity"—the purest form of scientific truth.[51]

One can compare McClintock's process to that of lovers and poets or to the masculine transcendence experienced by Einstein in his day. Indeed Keller liked to think of it as science that transcended—a science that transcended gender. It was kinder, less autocratic, but not more easily achieved by either sex, since McClintock, its practitioner, defied distinction as a gendered type. This brand of science was antihierarchical and democratic; anyone with humility could participate. Keller summoned it as proof that McClintock was not elitist or haughty, as colleagues claimed she could be toward people of lesser intelligence. Although she gravitated toward smart people, they were gentle people, often women, but not always. Her field of corn became a great social equalizer since it did not cultivate the power relationships of the lab. McClintock had no assistants to rule over or graduate students to evaluate; she planted her own corn, and when she was away she found high school girls more effective surrogate caretakers than the world's leading cytogenetic experts.[52]

McClintock was never a vocal advocate for women scientists, but Keller believed that her brand of science gave women tools, perspective, and confidence to trust their intuition. It wasn't McClintock's style to tell younger researchers what to think; rather, she would guide them to clearer visions of their thoughts. A female researcher remembered sitting in the living room of the old dorm at Cold Spring Harbor asking McClintock to help her link up the facts of her most recent experiments. McClintock told her the heretical: that inspired science was to some extent an imaginative act. The young bacterial geneticist Evelyn Witkin was inspired in the 1940s when McClintock told her not to worry about what textbooks told her to find under the microscope: don't tell nature how to be; rather, let nature "tell you where to go." In masculine science, intuition and intimacy were dirty words, but McClintock feared neither and had been well served by both.[53]

Listening to McClintock describe her process, Keller could see the irony. "Things are more marvelous than the scientific method allows us to conceive," McClintock explained, and yet she felt "in closer touch with biological reality" as colleagues thought her "increasingly out of touch." She was a self-proclaimed mystic who studied Buddhism, acupuncture, biofeedback, and other means of knowing that were at odds with Western science. Her seeming fascination with the supernatural didn't endear geneticists to her ideas in the 1950s and 1960s, though in later decades many of Keller's readers thought her interests romantically "primitive" and stereotypically "feminine." Witkin insisted that McClintock didn't believe in UFOs or ESP, but believed one shouldn't disbelieve without dispositive proof. Her mysticism was simply an admission of nature's complexity, that there was no way to understand all its mysteries. As a younger colleague put it, McClintock had "the courage to say, 'I do not understand.'"[54]

These were the seeds of McClintock's feminist mythos; they germinated through Keller's telling of events, which first occurred at the infamous Cold Spring Harbor Symposium in 1951. McClintock was nervous about presenting her research, for geneticists had long taken for granted that genes lined up like pearls on a string, stable and impervious to environmental stimuli. After years in her cornfields, McClintock was going to suggest something different, that genetic elements moved from one site to another, even between chromosomes, and that this movement was programmed by a complex system of controls influenced by their interactions with the environment. The jumping around of genetic elements, a phenomenon that came to be known as "transposition," was the discovery for which McClintock won her Nobel Prize thirty years later, but the majority of prokaryote geneticists she spoke to that day in 1951 were reluctant to see heuristic value in the controlling concept that lay at the heart of her theory. They were using radioisotopes, X-ray crystallography, and electron microscopes to study the molecular structure of simpler organisms, while McClintock used low-tech microscopes and crossbred her corn by hand. Rather than appear revolutionary, she seemed to prove that the age of maize genetics had come and gone while others passed her by.[55]

There was silence when McClintock finished her paper; she had confounded her peers, but Keller posited that she had likely also threatened them as arbiters of scientific truth. McClintock was so hurt by the response that she closed herself off from scientists in her field, refusing to publish or speak of her ideas to anyone except an inner sanctum of friends who understood. She stuck to her convictions, however, and colleagues continued to think her "mad" until the late 1960s and 1970s, when molecular biologists confirmed dynamism in the genomes of *E. coli* and other organisms. By the mid-1970s the women's movement was at its most vibrant, and other biographers were reexamining the scientific accomplishments of women obscured by history. Keller thought McClintock's story significant, less for the vindication she now enjoyed than for her unacknowledged feeling for organisms. Like feminist scholars since, Keller was interested in process more than scientific results.[56]

McClintock won the Nobel Prize in 1983, five months after Keller's biography, *A Feeling for the Organism*, was published. The prize brought readers to her story, but it also obscured the nuanced points Keller raised about gender and science. Readers who looked for a martyr of the male power structure saw one in McClintock; those who looked for a heroine among a female breed of scientists also saw what they wanted to see. Some thought her a female maverick, using male license; for others the moral of her story was that all, including men, could listen patiently—womanly—for nature to speak. In both cases, she became a figure almost exactly in opposition to the popularly imagined Marie Curie of 1920 or the Maria Mayer heralded in 1963. McClintock was a new idealized image of a female scientist not because of her motherhood or domesticity, but because of the intuitive way she worked at science. In this respect, Keller's aim to *regender* the portrait of a scientist had been successful.

After McClintock's death in 1992, historian Nathaniel Comfort looked through the transcripts of Keller's interviews and other archival evidence in search of the authentic McClintock. He found a woman of "intimidating intellect," "biting humor," and "fierce independence," but not necessarily a mystic or maverick or marginalized martyr. She claimed that her peers had rejected her views in 1951, and yet he discovered (and several scientists have since confirmed) that they discussed her ideas at length and generally accepted transposition in corn.[57] Comfort reiterated that McClintock was the third woman to be admitted into the American Academy of Science and the first to be president of the American Genetics Society. Honors had continued to come her way in the 1950s and 1960s, decades in which Keller claimed she had been largely ignored. By the early 1980s she had won the Thomas Hunt Morgan Medal, the Wolf Prize, and the Lasker Award, and became the first MacArthur laureate before winning the Nobel. Most scientists doubted her theories about controlling elements, and rightly so, Comfort believed, since they were never empirically confirmed. Transposition had never been the thrust of her work, and yet younger scientists generously remembered it differently. This was not a woman undermined, he determined; he couldn't see the lasting impact of sexism in her career. If her tale of victimization was a sham, so was the idea that McClintock adhered to precepts of reason that were different from her peers'. Comfort called her process "integration" and claimed that, since it had also been Einstein's and Feynman's, there was nothing mystical or female about it.[58]

So then why the myth? Comfort believes timing was a factor. McClintock was a Nobel Prize winner who needed an appealing narrative, and maverick behavior, mysticism, and a healthy dose of martyrdom would do. James Watson had once referred to the trousered, temperamental McClintock as the "Katharine Hepburn of science," but only in her eighties was the characterization a compliment. "She proved an irresistible media figure," Comfort explained, "the boyish little old lady, dragged out of the solace of the laboratory, blinking up at the klieg lights, saying everyone had thought she was crazy, she knew all along she was right, she hated publicity, had no need for money."[59] Keller saw her differently, revealing the contested nature of the woman scientist as cultural icon. Thomas Kuhn understood that scientists unearthed socially constructed truths, but of course historians and biographers do too.

The "transformation" suggested by the title of this section was Keller's as a feminist and McClintock's as feminist legend, but it was also a singular transformation made manifest through both women. Keller's feminism changed her way of thinking about scientific inquiry, and McClintock became her vehicle for showing this scientific inquiry embodied and practiced. Where her colleagues looked for rules in nature, McClintock believed that there was value in studying anomaly for anomaly's sake. For Keller, McClintock was that anomaly, that mutation, that exception to many rules, and so she was worth getting to know on her terms. Keller's time with McClintock confirmed that everything was indeed connected. McClintock's eccentricities as a woman and a scientist gave Keller insight into whole systems of gender and science as well as insight into herself. Keller achieved her own feeling for the organism.[60]

Notes

1. Taken from the tenth-anniversary edition (Yale University Press, 1995), 174–75.

2. Elizabeth Stone, "A Mme. Curie from the Bronx," New York Times Magazine, April 9, 1978, 29; Sharon Bertsch McGrayne, Nobel Prize Women in Science: Their Lives, Struggles, and Momentous Discoveries, 2nd ed., (Washington, DC: Joseph Henry Press, 1998), 336.

3. Rosalyn Yalow, "Autobiography," in The Nobel Prizes, 1977, ed. William Odelberg (Stockholm: Nobel Foundation, 1978); Joan Dash, The Triumph of Discovery: Women Scientists Who Won the Nobel Prize (Englewood Cliffs, NJ: Julian Messner, 1991), 38–42; Olga S. Opfell, The Lady Laureates: Women Who Have Won the Nobel Prize (Metuchen, NJ: Scarecrow Press, 1986), 261; Diana C. Gleasner, Breakthrough: Women in Science (New York: Walker, 1983), 47; Nancy J. Veglahn, Women Scientists (New York: Facts on File, 1991), 106.

4. Adolph Friedman, "Remembrance: The Berson and Yalow Saga," Journal of Clinical Endocrinology and Metabolism 87, no. 5 (2002): 1925–28; Eugene Straus, Nobel Laureate Rosalyn Yalow: Her Life and Work in Medicine (Cambridge, MA: Perseus Books, 1998), 33–34; Yalow, "Autobiography"; McGrayne, Nobel Prize Women, 337–38; Veglahn, Women Scientists, 108.

5. Yalow, "Autobiography"; Straus, Nobel Laureate Rosalyn Yalow, 36–37, 56, 103, 168, 176; Dash, Triumph of Discovery, 44–47.

6. Yalow apparently didn't know that Quimby was a mother or perhaps forgot, since in later years she questioned her mentor's decision to remain childless. See Straus, *Nobel Laureate Rosalyn Yalow*, 68, 106; Veglahn, *Women Scientists*, 48–54; Mildred Dresselhaus, "Rosalyn Yalow," in *Out of the Shadows: Contributions of Twentieth-Century Women to Physics*, ed. Nina Byers and Gary Williams (New York: Cambridge University Press, 2006), 308–309; Dash, *Triumph of Discovery*, 48.

7. Barbara Shiels, *Winners: Women and the Nobel Prize* (Minneapolis: Dillon Press, 1985), 47, 51–52; McGrayne, *Nobel Prize Women in Science*, 336; Straus, *Nobel Laureate Rosalyn Yalow*, 83.

8. Straus, *Nobel Laureate Rosalyn Yalow*, 157–58; Dash, *Triumph of Discovery*, 51–52.

9. Straus, *Nobel Laureate Rosalyn Yalow*, 164–75; Mary Harrington Hall, "The Nobel Genius," Box 9, folder 30, Maria Goeppert Mayer Collection (MGM).

10. At the Nobel banquet, Yalow said she "[gave] birth to and nurtured through its infancy radioimmunoassay, a powerful tool for the determination of virtually any substance of biologic interest." See Straus, *Nobel Laureate Rosalyn Yalow*, 112; McGrayne, *Nobel Prize Women in Science*, 341; Gleasner, *Breakthrough*, 47, 53; Shiels, *Winners*, 63–64; Opfell, *Lady Laureates*, 256, 263; Yalow, "Autobiography."

11. Friedman, "Remembrance"; Straus, *Nobel Laureate Rosalyn Yalow*, 79–80, 170, 174–77, 223–24; McGrayne, *Nobel Prize Women in Science*, 342.

12. Helena M. Pycior, "Pierre Curie and 'His Eminent Collaborator Mme Curie,'" in *Creative Couples in the Sciences*, ed. Helena M. Pycior, Nancy M. Slack, and Pnina G. Abir-Am (New Brunswick, NJ: Rutgers University Press, 1996), 39–56; Shiels, *Winners*, 54–55; Dash, *Triumph of Discovery*, 50–52; Straus, *Rosalyn Yalow*, 170, 177, 231–32; McGrayne, *Nobel Prize Women in Science*, 342; Gleasner, *Breakthrough*, 44–46.

13. Friedman, "Remembrance"; Yalow, "Autobiography"; Dash, *Triumph of Discovery*, 62.

14. McGrayne, *Nobel Prize Women in Science*, 350; Straus, *Nobel Laureate Rosalyn Yalow*, 236–37.

15. Shiels, *Winners*, 60; Straus, *Nobel Laureate Rosalyn Yalow*, 239–41; Gleasner, *Breakthrough*, 33.

16. Veglahn, *Women Scientists*, 107; Straus, *Nobel Laureate Rosalyn Yalow*, 78, 81; McGrayne, *Nobel Prize Women in Science*, 337, 354; Gleasner, *Breakthrough*, 51; Shiels, *Winners*, 63–64.

17. Straus, *Nobel Laureate Rosalyn Yalow*, 107, 171, 247–57; McGrayne, *Nobel Prize Women in Science*, 336, 340, 353.

18. Most prominent of the studies enabling the "superwoman" ideal was Jonathan Cole and Harriet Zuckerman, "Marriage, Motherhood, and Research Performance in Science" (1987), in *The Outer Circle: Women in the Scientific Community* (New Haven, CT: Yale University Press, 1992). See also Ann Gibbons, "Key Issue: Two-Career Science Marriage," *Science* 255 (March 13, 1992): 1380.

19. Joan Freeman, *A Passion for Physics: The Story of a Woman Physicist* (Bristol, U.K.: Adam Hilger, 1991). An autobiography with the same sensibilities is Rita Levi-Montalcini, *In Praise of Imperfection: My Life and Work* (New York: Basic Books, 1988).

20. William C. Kelly to Maria Mayer, July 28, 1964, Box 10, folder 19; "Can Working Woman Become Good Wife"; "Nobel Prize Winner Has Sound Advice for Women," *Mainichi Evening News*, April 9, 1965, Box 9, folder 31, MGM.

21. Alice Rossi, "Barriers to the Career Choice of Engineering, Medicine, or Science Among American Women," in *Women and the Scientific Professions: The MIT Symposium on American Women in Science and Engineering*, ed. Carol G. Van Aken and Jacquelyn A. Mattfeld (Cambridge, MA: MIT Press, 1965), 125–27.

22. Margaret W. Rossiter, *Women Scientists in America: Before Affirmative Action, 1940–1972* (Baltimore: Johns Hopkins University Press, 1995), 41–42.

23. See, for example, "Two Americans and German Get Nobel Physics Prize," *New York Times*, Western edition, November 6, 1963; "U.S. Shares Nobel Prize; California Woman Wins," *Washington Evening Star*, November 5, 1963; "12th Nobel Prize for U.C. Faculty," *San Diego Evening Tribune*, November 6, 1963, 42, Box 9, folder 28; Esther Gwynne, "Nobel Winner Home Again," *San Diego Evening Tribune*, December 24, 1963, Box 9, folder 29, MGM.

24. Mary Harrington Hall, "American Mother and the Nobel Prize—a Cinderella Story in Science," *McCall's*, July 1964; "The Nobel Genius," Box 9, folder 30, MGM.

25. Maria Mayer to Robert L. Weber, June 11, 1969, Box 2, folder 9; Hall, "Nobel Genius," MGM.

26. Maria Mayer to Mary Markley, April 20, 1964, Box 2, folder 3, MGM.

27. Vera Kistiakowsky to Maria Mayer, April 16, 1971, Box 2, folder 11; Maria Mayer to Rita Arditti and Elisa Buonaventura, December 10, 1970, Box 2, folder 10, MGM.

28. Vera Rubin, "Cecilia Payne-Gaposchkin," in *Out of the Shadows*, 167; Elga Wasserman, *The Door in the Dream: Conversations with Eminent Women in Science* (Washington, DC: Joseph Henry Press, 2002), 42; Levi-Montalcini, *In Praise of Imperfection*; Joan Dash, *Triumph of Discovery*, 99–130.

29. Maria Goeppert Mayer, "Manuscript Changes," n.d., Box 9, folder 2, MGM; Joan Dash, *A Life of One's Own: Three Gifted Women and the Men They Married* (New York: Paragon House, 1988), x.

30. Brenda Maddox, *The Dark Lady of DNA* (New York: HarperCollins, 2002), 65, 93; Anne Sayre, *Rosalind Franklin and DNA* (New York: W. W. Norton, 1975); McGrayne, *Nobel Prize Women in Science*, 303–31; Francis Crick, *What Mad Pursuit: A Personal View of Scientific Discovery* (New York: Basic Books, 1988), 88.

31. Horace Freeland Judson, *Eighth Day of Creation: The Makers of the Revolution in Biology* (New York: Simon and Schuster, 1979), 148–49; Philip Siekevitz to Dr. Aaron Klug, March 26, 1976; A. Klug to Dr. P. Siekevitz, April 14, 1976, Rosalind Franklin Papers, Churchill Archives, Churchill College, Cambridge, U.K. (letter digitized in cooperation with the U.S. National Library of Medicine, Digital Manuscripts Project, Bethesda, MD).

32. Straus, *Nobel Laureate Rosalyn Yalow*, 167.

33. Vivian Gornick, *Women in Science: Then and Now* (New York: The Feminist Press, 2009), 106; *Women in Science: Portraits from a World in Transition* (New York: Simon and Schuster, 1983), 120.

34. Fay Ajzenberg-Selove, *A Matter of Choices: Memoirs of a Female Physicist* (New Brunswick, NJ: Rutgers University Press, 1994), 3.

35. Ajzenberg-Selove, *A Matter of Choices*, 33, 49, 54, 73–74, 80, 83–84.

36. Ajzenberg-Selove, *A Matter of Choices*, 114–15, 159–61.

37. Ajzenberg-Selove, *A Matter of Choices*, 162–66.

38. Chien-Shiung Wu, "The Commitment Required of a Woman Entering a Scientific Profession," in *Women and the Scientific Professions*, 44–48; Iris Noble, *Contemporary Women Scientists of America* (New York: Julian Messner, 1979), 82; McGrayne, *Nobel Prize Women*, 266–75; Moira Reynolds, *American Women Scientists: 23 Inspiring Biographies, 1900–2000* (Jefferson, NC: McFarland, Inc., 1999), 116.

39. Mildred S. Dresselhaus, "Responsibilities of Women Faculty in Engineering Schools," in *Women in the Scientific and Engineering Professions*, ed. Violet B. Haas and Carolyn C. Perrucci (Ann Arbor: University of Michigan Press, 1984), 129–32; "Perspectives on the Presidency of the American Physical Society," *Physics Today*, July 1985, 37–44;

"Women Graduate Students," *Physics Today*, June 1986, 74–75; G. Dresselhaus, "Mildred Spiewak Dresselhaus," in *Out of the Shadows*, 355–61; Noble, *Contemporary Women Scientists*, 143–51; Straus, *Nobel Laureate Rosalyn Yalow*, 162.

40. Elisabeth J. Fraenkel to Vera Kistiakowsky, December 8, 1971, Box 2, folder 11, MGM.

41. Gornick, *Women in Science* (2009), 79–80; Wasserman, *Door in the Dream*, 4–6.

42. Gornick, *Women in Science*, 63, 80, 82, 86–87, 88–89.

43. Gornick, *Women in Science*, 61, 76, 107, 116, 128, 133–38.

44. Evelyn Fox Keller related her experience as a Harvard graduate student and her transition into the philosophy of science in Sara Ruddick and Pamela Davis, eds., *Working it Out: 23 Women Writers, Artists, Scientists, and Scholars Talk About Their Lives and Work* (New York: Pantheon Books, 1977), 78–91. See also Andrew Brown, "Fox Among the Lab Rats," *Guardian*, November 4, 2000 (online archive) (http://www.guardian.co.uk/books/2000/nov/04/books.guardianreview6).

45. Evelyn Fox Keller, *Reflections on Gender and Science* (New Haven, NJ: Yale University Press, 1995), 3–5; *Refiguring Life: Metaphors of Twentieth-Century Biology* (New York: Columbia University Press, 1995), ix; Brown, "Fox Among the Lab Rats."

46. Evelyn Fox Keller, *A Feeling for the Organism: The Life and Work of Barbara McClintock* (New York: Henry Holt, 1983), 83–84; *Reflections on Gender and Science*, 158, 173.

47. Keller, *A Feeling for the Organism*, 83–86; Dash, *Triumph of Discovery*, 75–85; Lee B. Kass, "Records and Recollections: A New Look at Barbara McClintock, Nobel Prize-Winning Geneticist," *Genetics* 164 (August 2003): 1251–60.

48. Barbara McClintock to Charles Burnham, February 20, 1930; April 2, 1935; September 16, 1940; October 9, 1940, Correspondence, Barbara McClintock Papers, National Library of Medicine, Rockville, MD.

49. Videotaped interview of Evelyn Witkin, Barbara McClintock Audio/Video Archive, Cold Spring Harbor Laboratory, Digital Archives, http://library.cshl.edu/mcclintock/mcclintock_av.html; Charles D. Laird, "The Plural of Heterochromatin," in *The Dynamic Genome: Barbara McClintock's Ideas in the Century of Genetics*, ed. Nina Federoff and David Botstein (Plainview, NY: Cold Spring Harbor Laboratory Press, 1992), 155; Keller, *A Feeling for the Organism*, 17–34, 198.

50. Keller, *A Feeling for the Organism*, 101, 207, 104; Bruce Alberts, "Please Come to My Laboratory for Better Coffee, Fresh Orange Juice, . . . Conversation," in *The Dynamic Genome*, 279.

51. Keller, *A Feeling for the Organism*, 117; Lisa Heldke, "John Dewey and Evelyn Fox Keller: A Shared Epistemological Tradition," in *Feminism and Science*, ed. Nancy Tuana (Bloomington: Indiana University Press, 1989), 112–13.

52. Barbara McClintock to Milislav Demerec, February 20, 1953, Box 1, folder 5, Barbara McClintock Digital Archive, Cold Spring Harbor Laboratory Archives (CSHL).

53. Keller, *A Feeling for the Organism*, 125, 203–4; Lilla Fano to Barbara McClintock, January 8, 1972, Box 1, folder 6, Barbara McClintock Papers, National Library of Medicine; Evelyn Witkin, "Cold Spring Harbor 1944–1955: A Minimemoir," in *The Dynamic Genome*, 116.

54. Keller, *A Feeling for the Organism*, 180, 192–93; James Shapiro, "Kernels and Colonies: The Challenge of Pattern," in *The Dynamic Genome*, 216; Nathanial C. Comfort, *The Tangled Field: Barbara McClintock's Search for the Patterns of Genetic Control* (Cambridge, MA: Harvard University Press, 2001), 153.

55. Mel Green, "Annals of Mobile DNA Elements in Drosophila: The Impact and Influence of Barbara McClintock," in *The Dynamic Genome*, 117–22; Dash, *Triumph of Discovery*, 86–87.

56. Barbara McClintock to Dr. J. R. S. Fincham, May 16, 1973, Barbara McClintock Papers, National Library of Medicine; Keller, *A Feeling for the Organism*, 137–42.

57. Green, ."Annals of Mobile DNA Elements in Drosophila," 117–22; Ira Herskowitz, "Controlling Elements, Mutable Alleles, and Mating-type Interconversion," 289–97; Nina Fedoroff, "Maize Transposable Elements: A Story in Four Parts," 389–415, in *The Dynamic Genome*.

58. Comfort, *The Tangled Field*, 1–16, 32–68; Keller, *A Feeling for the Organism*, 10.

59. Comfort, *The Tangled Field*, 246–47.

60. Heldke, "John Dewey and Evelyn Fox Keller," 104–15; Keller, *A Feeling for the Organism*, xxii.

7 The Lady Trimates and Feminist Science?: Jane Goodall, Dian Fossey, and Biruté Galdikas

We think of science as manipulation, experiment, and quantification done by men dressed in white coats, twirling buttons and watching dials in laboratories. When we read about a woman who gives funny names to chimpanzees and then follows them into the bush, meticulously recording their every grunt and groom, we are reluctant to admit such activity into the big leagues. We may admire Goodall's courage, fortitude, and patience but wonder if she represents forefront science or a dying gasp from the old world of romantic exploration.... The conventional stereotype is so wrong. ... Jane Goodall's work with chimpanzees represents one of the Western world's great scientific achievements.

—Stephen Jay Gould, Introduction to the
revised edition of *In the Shadow of Man*[1]

Often I think of science in technological terms—of the cold machinery, the devices, and accelerators, the weapons that science makes possible— all the things that modern science creates and utilizes. However, one day, I thought of science and appreciated its intent to look more closely into the beauty and mystery of nature. I had a glimpse of science in a different light, and at that moment the image of the woman in my dream came to mind. In one view of science the image exists of the male scientist exerting power and control over passive female nature. In this view the practice of science is seen as a violation of the natural world. However, my dream image raised the possibility of an alternative view. I began to consider another generative impulse of pure science—one born of curiosity and the love of nature. Then the woman becomes an intriguing symbol of a new way for me to think about the practice of science and its nature. She embodies the sense of science as the desire to understand nature, pursued in a rational and imaginative way. . . . Science is then not about the power of (male) intellect over passive (female) embodied nature. Rather science is a marriage, the relationship between human intellect and the intelligibility of a dynamic nature—nature which is both mysterious and knowable and in whose knowing we learn something about ourselves.

—Mary Palevsky, *Atomic Fragments*, April 1997[2]

IN 1982 THREE WOMEN CONVENED AT AN EXPLORERS CLUB RECEPTION in NEW York City. From outward appearances, they were in their thirties and forties. One was a dulling blond, the other two brunettes; one towered over the other two, who appeared to be of average height. Each wore an informal print cotton dress, though one filled hers out with a clearly pregnant belly. From the pleasant, relaxed expressions on their faces, one might have thought they were wives of important scientists being honored for their research in the field, when in fact the three women were themselves the most famous experts in the world on wild primates. Men in universities on nearly every continent had read as much, if not more, than they on chimpanzees, gorillas, and orangutans; but no one had logged more than a fraction of their hours in live observation of the animals in their natural habitats. Collectively, the women had spent more than forty years in the forest, and they had only just begun.

Jane Goodall, the first of this scientific triumvirate, was a soft-spoken Englishwoman who had studied the chimpanzees of the Gombe River Reserve of Tanzania since 1960. A forty-eight-year-old divorcée and recent widow, she had lost some of the youthful glow that had drawn the attention of more than three million people to pictures of her in National Geographic in 1963. Nearly twenty years later, she was still slim and dignified, her hair tied back in the same loose ponytail. Initially, academics had written her off as a National Geographic cover girl. She was referred to in newspaper articles as the blonde who "preferred chimps to men," but her life in Africa had not been a publicity stunt after all.

Throughout her girlhood, animals had consumed her interest, and she became inexhaustibly curious and patient observing them. At the age of four she waited for hours to witness chickens laying eggs; by nine she was riding horses, by eleven drawing pictures highlighting the differences between green loopers and caterpillars that turned into lime hawk moths. When most adolescent girls were writing to friends about high school crushes, she sat with her nature log classifying bullfinches and hedge sparrow babies. Her fascination with animals did not lead to formal study at university. Instead she held waitress and secretarial jobs until she had the money in hand to fly to Kenya. The twenty-three-year-old found work in Nairobi as a secretary at the Coryndon Museum, under its illustrious director, paleontologist Louis Leakey. After taking her to assist his wife at their dig site at Olduvai Gorge, he chose Goodall to carry out a long-term study of chimpanzees in Tanzania.[3]

The chair of the National Geographic Society, the organization that became her greatest sponsor, remembered what he thought of the novice explorer when he first met her: "She was obviously bright, but in-experienced, with a high-school equivalent education. . . . I thought it very unlikely this attractive young woman would devote her life to studying chimpanzees at Gombe Stream, deep in the wild, remote for-est of Tanzania." And yet she agreed to go into "the wild," a five days' drive from Nairobi and twelve miles from the nearest town of Kigoma. Only some local fishermen, a native cook, and her mother, Vanne Goodall, accompanied her, since Tanzanian officials refused to let her proceed without a parental chaperone. In the first months at Gombe, she was bedridden with high fever and slowed by malaria, the nearest doctor three hours away by boat. She turned waif thin as she adjusted to the heat and a reduced intake of calories, but she eventually felt well enough to find a lookout point high above the valley. Coffeepot in tow and notebook in hand, she waited patiently for the chimpanzees to make an appearance, even if it meant wrapping blankets around her to endure the night air.[4]

Perched at her post, in October 1960 she observed chimps inserting long twigs into termite mounds to scrape for food; a week later she noted their consumption of baboon babies, bushbuck, and several kinds of monkey. Little did she know that, as she recorded these behaviors, she was cracking evolutionary theory from its foundations, challenging the definition of *Homo sapien* in distinction from other species. The world's leading primatologists had long insisted that chimps were vegetarian and couldn't use tools, but this relative novice quickly proved them wrong. Soon she also reported on the chimps' travel routes and child-rearing practices and claimed to understand the organization of their social structure. Specialists were intrigued by her fortuitous findings and came to see for themselves. By 1975 the Gombe Stream Research Center teemed with graduate students and scholars from Stanford, Cambridge, and the University of Dar es Salaam. Goodall had created a truly international center of research.

In 1986 many of the scholars who once made up the Gombe com-munity came to Chicago to hold a conference marking the publication of *The Chimpanzees of Gombe: Patterns of Behavior*, the most compre-hensive compendium of Goodall's research to date. By this time she was no longer merely a leading light in academic circles, but a darling of the American public. Since 1977 an institute bearing her name had advocated for habitat conservation and better treatment of animals in captivity. Promotional trips to the United States soon required personal assistants to organize her book signings and television spots. By the 1990s Jane Fonda, Johnny Carson, Jimmy Stewart, and Michael Jackson were helping her raise funds for the apes. Her books for laypeople and children turned her into one of the most popular science writers in the world. No man or woman had translated scientific research into accessible language and advocacy so successfully since Rachel Carson wrote on the perils of pesticides in 1962.[5]

Jane Goodall was a tough act to follow when Louis Leakey's second "Lady Trimate," Dian Fossey, embarked on her study of mountain gorillas in the Congo in 1966. A six-foot-two American, she was rougher around the edges than her reserved British counterpart. Fossey looked composed for photographers at the Explorers Club meeting in 1982, but she was anxious to get back to her animals in the Virunga Mountains. It was no secret that she liked gorillas much better than people. Brash, awkward, and, at times, downright hostile to Westerners, she increasingly preferred her own company in her isolated cabin at Camp Karisoke to engaging the egos and curiosities of those from outside.[6]

Dian Fossey grew up in an upper-middle-class family in San Francisco and had been estranged from her mother and stepfather since she was a child; animals had long become surrogates for affection. In high school she rode show horses and decided that her career would be one of caring for them, if only she could make the grades in science. She struggled in chemistry too much to get into a veterinary program and settled instead for courses in animal husbandry, eventually majoring in the occupational therapy of children, but she never relinquished her fantasy of going to Africa to study animals up close. When she read about Jane Goodall in 1963, she took out a loan against three years' salary and set out for Tanzania, hoping that Louis Leakey would lead her to the animals. He put her in contact with photographers who allowed her to accompany them at Kabara, the study site of American gorilla expert George Schaller just three years earlier. It was here that she became enchanted with the mountain gorilla, but also where she first detected her physical limitations. Within seven weeks she had been so debilitated by ankle sprains that she was forced to return to the United States, her dream of studying animals apparently over.

Fossey wrote stories about the gorillas for local newspapers and got engaged to a Rhodesian man who had come to the United States to study and had no intention of returning home. Meanwhile, Louis Leakey was on an American lecture tour and came across her stories and photographs. He asked her if she wanted to study gorillas just as Jane Goodall studied chimps, and her response was immediate. She tabled her wedding and walked away from a career in physical therapy to embark on a life in the field, telling her family and fiancé that she didn't know when she'd be back.[7]

Fossey left her life behind to study a group of gorillas that occupied an area of extinct volcanoes that stretched for twenty-five miles, two-thirds of which belonged to the Democratic Republic of the Congo, and the rest situated in Rwanda's Parc National des Volcans and Uganda's Kigezi Gorilla Sanctuary. It was the site on which legendary naturalist Carl Akeley had convinced King Albert of Belgium to create a national park that would protect gorilla inhabitants. Setting up camp four thousand feet up, Fossey began her life as an observer and ally of mountain gorillas. When civil war broke out in the Congo, rebel soldiers put her under house arrest, but she convinced them to drive her into Uganda. Reports differ about what happened across the border. She told some that she was brutally raped, others that she endured various forms of torture. She emerged from the incident determined to set up a new camp on the Rwandan side of the Virunga Mountains, a camp that she named Karisoke in September 1967.[8]

Ten thousand feet up from the closest neighboring villages, she hired porters to trek up and down the mud-drenched, nettle-laden mountain twice a week to bring food and mail to her camp. Men trained for this work could do it in an hour and a half; with her arthritic legs and emphysemic lungs, she needed the better part of four hours. It's unlikely that a day went by on the mountain that she wasn't in pain. She suffered hemorrhages, near blindness, rotting teeth, broken bones, and diseases likely ranging from tuberculosis to cancer. Before she arrived in Africa she had weak lungs, made worse by the thin air and her smoking of unfiltered cigarettes. The scarcity of fair weather, fresh food, and friends did little to improve her condition. "It was indeed fortunate that I really liked potatoes," she joked, for sometimes the only choice was mashed, baked, or boiled. Students who came to study the gorillas often left as quickly as they arrived. She thought it just as well: "It never dawned on me that exhausting climbs along ribbons of muddy trail, bedding down in damp sleeping bags, awakening to don wet jeans and soggy boots, and filling up on stale crackers would not be everyone's idea of heaven."[9]

Karisoke was not convenient, but since it was the best place from which to study rare mountain gorillas, that's where she stayed. Her sub-species of interest, *Gorilla gorilla beringei*, had been "discovered" in 1902, but never studied closely, and, Fossey speculated, might be extinct before the end of the century. She was determined to understand and help the gorillas, her attitude raising perpetual conflict with local cattle-men and poachers. Hostilities between Bahutu and Batutsi farmers had long led to the illegal grazing of cattle in zones designated for the goril-las, and Batwa hunters poached gorillas to make novelty items of their hands, feet, and heads. Government officials and researchers wanted to protect the animals and hoped that a regular flow of visitors would bring wealth into the local economy and provide an incentive that would allow for such protection. But Fossey didn't think there was time to wait for the benefits of tourism to kick in—that is, if the tourists themselves did not lead to the destruction of the animals. Regular contact with hu-mans could make the gorillas more vulnerable to diseases and traps. She believed that patrols, if their members were provided adequate salaries and weapons, would pose less of a threat to the animals.[10]

No scientist, farmer, or government official was interested in her ap-proach, so she carried out vigilante justice herself. Leveraging research-ers and local tribesmen, she instructed them to cut traps, confiscate weapons, and release animals from snares, regardless of the local ten-sions they incited. If graduate students wanted to see the animals, they were going to have to take on the dangerous patrolling responsibilities too. Many of them, fresh from the lab, proved unfit for the challenge and watched in horror as she staged sacrificial rituals to scare the lo-cals into compliance. Their belief in black magic proved useful; during trips back home she replenished her supply of Halloween masks for her charades. Researchers reported that she whipped genitals and injected poachers with gorilla dung until septicemia set in. Allegedly, she kid-napped a poacher's son to make him submit. All of it, she insisted, was for "the sake of the remaining gorillas."[11]

Corporate sponsors were not impressed. Whereas Jane Goodall re-mained the success story of the National Geographic Society (NGS) and the Wilkie and Leakey Foundations, Fossey became persona non grata in the community of primatologists. Like Goodall, she had come to Africa in search of apes, steady funding, professional affirmation, and perhaps even companionship; but Goodall had been more successful on every count. Goodall's work in Tanzania led her to the men she eventually married: NGS photographer Hugo van Lawick in 1964 and Tanzanian parks director Derek Bryceson in 1975. With van Lawick, Goodall bore a beautiful blond son, known as Grub, much to the fascination of readers who followed the event through the glossy images of *National Geographic*. Life and love were never as easy for Fossey. After years of grueling effort in unbearable conditions, physical contact with the apes was elusive. For years she had to settle for a brief touch of fingertips with the great silverback Peanuts. The pictures that photographer Bob Campbell was able to capture of the event were not nearly as flattering of his human subject as van Lawick's were of his. Fossey, too, fell in love with her photographer, but Campbell left her and went back to his wife. He would not be the last married man to choose his wife over her. Although she wanted children, her unsuccessful love affairs resulted in dangerous abortions performed without the benefits of Western medical equipment. Not a day went by that she did not fear that financial hardships would bring her research to an end. She was bitter and often lashed out at staff and colleagues.[12]

Rwandan locals referred to Fossey as *Nyiramachabelli*, which translates to something like "the woman who lives alone on the mountain." (A British tabloid translated it to mean "the old lady who lives in the forest without a man.") Ann Pierce, who spent fourteen months at Karisoke, described it as "very lonely up there"; students unprepared for the isolation exhibited symptoms of "astronaut blues," a combination of "sweating, uncontrollable shaking, short-term fevers, loss of appetite, and severe depression combined with prolonged crying spells." If students felt this way after only brief stints, one can imagine the extent to which eighteen years at Karisoke plagued Fossey's psyche. Accounts differ, but most corroborate that she grew frighteningly violent and mad. In 1977 the brutal killing of Digit, her most beloved gorilla, seemed to send her over the edge. She was murdered in her bed in December 1985. Two men were convicted of the crime, but in truth no one really knows who killed her, for she had many enemies around her.[13]

The third woman of Leakey's triumvirate, Biruté Galdikas, headed not for Africa but for the rainforests of Kalimantan, Indonesian Borneo. Fossey quarreled with her, never realizing that her fellow "Trimate" would defend Fossey's militant protection of animals more eloquently than anyone else. When others questioned Fossey's sanity, Galdikas credited her with single-handedly converting the public image of the gorilla from "monstrous King Kong to peaceful vegetarian" and for ultimately saving him from extinction.[14]

In high school Galdikas dreamed of going to the Far East to be near the orangutans, but unlike the other Lady Trimates, she had acquired a husband and nearly her doctoral credentials before embarking into the field. She met Louis Leakey during yet another of his American tours: he had just given a lecture to her archaeological-dating-techniques class at the University of California, Los Angeles, when she approached him. His walking cane and nearly toothless smile belied his vigor; his enthusiasm was infectious—nearly "evangelical," as she experienced it. She declared that she was his next lady protégée; she had already written the Malaysian government and planned to be accompanied to the rainforest by her husband, physicist Rod Brindamour. He would double as her photographer and camp manager, much as Hugo van Lawick had for Goodall.[15]

Leakey helped her get funding from NGS and other private foundations, and she arrived in the dense swamps of Borneo in 1971. She named her camp after Leakey and embarked on her "return to nature" in the same idealistic spirit that permeated California campuses at the time. Fossey resented Galdikas's naive idealism as well as her charmed academic career and sex life, for although Galdikas worked in an oppressive rainforest, she seemed never to break a literal or figurative sweat. The pages of National Geographic displayed a long-haired beauty, who in no time had adapted to the swamps and had loving orangutan orphans hanging from her limbs. Galdikas admitted later that the appearance of Eden was only appearance: the rainforest was not a picnic site, and marriage in it was hardly idyllic. National Geographic had failed to capture her first week at Tanjung Puting, when she first wondered what she had gotten herself into. "I was confined to a tiny, airless hut with five men," she not so fondly recalled. "I had no privacy except after sundown, when oil lamps made from small sardine cans left most of the hut in darkness." The isolation was stifling. She longed for any reminders of her former existence—a letter, even an old magazine. "At times," she admitted, "we felt as though the rest of the world had abandoned us."[16]

Back home, friends and family feared that she and Rod were forever lost in the jungle, and the truth wasn't far off. For months they trekked with machetes in tow, clearing the thick growth that enveloped them. Avoiding fallen logs, vipers, and leech-infested pools was a treacherous and daily exercise. It was not uncommon for her to wade in swamp water to her armpits and to lose feeling in her frozen extremities. She layered her clothes to keep her notebooks dry but still succumbed to mysterious rashes, fevers, bites, and feelings of perpetual wetness. The orangutans, meanwhile, gave little indication of their presence in the forest other than an occasional brushing of leaves or shower of urine from the canopies overhead. By the time she saw her first animal in the wild, she had lost twenty-five pounds.[17]

In 1977 Galdikas became pregnant and, to the horror of her family, gave birth to her son without the benefits of modern medicine. She felt pressure to produce a respectable dissertation, and writing and mothering occupied all her time, while the native nanny occupied Rod Brindamour's. He no longer wanted to be with Galdikas, so she divorced him in 1979 as well as the Western existence she once knew. Essentially changing places with the nanny, Galdikas stayed in Indonesia and let her toddler son live with his father and teenaged wife in Canada. Later, she married a young Dayak tracker, a man smaller and less formally educated than the Western men she knew. He did not speak English, nor did he intend to leave the village where he was born. Their children would come to have Western names, Frederick and Jane, her daughter's name inspired by one of the few white women who understood the allure of the wild. But Galdikas eventually considered the Dayaks her people, the orangutans her family, and Borneo her permanent home.[18]

When she arrived in New York in 1982, Galdikas looked as though she had chosen a life of druidism over Western science or domesticity. She wore her signature oversized spectacles and was soon to give birth to another child. Her silky brown hair had become increasingly wild and soon would turn wiry gray. Once slim and voluptuous, with each passing year she gained weight and morphed more fully into a figure indifferent to Western tenets of beauty. Had she been a man, her transformation might have been likened to that of H. G. Wells's ingenious Dr. Moreau or of Conrad's Kurtz in *The Heart of Darkness*. She, too, had become a demagogue of sorts, assuming the identity of "Ibu," or "Mother Biruté," among the locals, but her succumbing to the wild seemed less romantic than for men. Shedding all the trappings of conventional life to live reclusively with her beloved animals, she appeared to be the proverbial cat lady; she didn't live in a dilapidated house down the street, but rather in a hut in the rainforests of Borneo—ultimately secluded from civilization.[19]

Galdikas transgressed the boundaries of female behavior but also of sound science, to the irritation of her professional peers. In the mid-1970s, primatologists had praised the rigor of her dissertation. Her findings on orangutan food sources supported revolutionary theories about divisions of labor that she extended to human males and females. After 1974 researchers adopted her animal sampling techniques as the most effective for recording behavioral data in the field. But like Fossey, she gradually fell from grace—down a slippery slope, from seeming to be a disinterested scientist to becoming an advocate for the apes. The rate at which she published research trailed off; thousands of pages of handwritten field notes sat in storage, collecting dust. Anthropologist Peter Rodman couldn't bring to mind any original research of hers after 1975. Carey Yeager, who observed proboscis monkeys upriver from Camp Leakey, believed that the ruse of her performing "science" was pretty much up when she started accepting tourists' money to play out fantasies of being scientists in the field. She directed more research in Bahasa Indonesian, rather than English, and admitted having little concern for remaining disinterested in the Western sense.[20]

Signs of her dramatic transformation were present even in 1975, as she lectured to a roomful of scientists in Los Angeles. Standing at the podium barefoot, she had appeared to have "gone native." They could not have known that she was not making a statement, but rather acting out of necessity. The swamps of Kalimantan had swollen her feet to such an extent that her closed-toe shoes didn't fit anymore. She had no need for formal footwear in Borneo and was surprised, when she returned to the West, to find how much her feet had expanded. Her feet are an ideal metaphor for her ideas about the natural world, a physical reflection of interior change that others might never understand unless they, too, had to walk in shoes that no longer fit. "Journal articles and monographs on fieldwork talk about theory, techniques and results," she reflected. "One rarely hears how fieldwork changes people's lives.

The living conditions, the funding difficulties, the practical problems, the highs of discovery, the false starts and dead ends, the drudgery of scientific record-keeping, the learning how to get along with people and societies initially very foreign to you, the learning how to get along without people, places, and things you once took for granted, the feeling of suspension in time as the world spins on without you—all have an impact."[21] Until Western scientists left the lab to live in the field, they couldn't know life in her shoes.

Louis Leakey's "Primitive" Feminism

The Trimates tracked uncharted paths toward communities of apes that were little understood by anyone. One can only wonder what each was thinking; men, let alone women, hadn't spent the intense time they would observing animals in their native habitats. Local officials thought Goodall insane for having no male escorts; colleagues thought Fossey suicidal when she set up camp with no survival training; and Galdikas traveled to areas so remote that her reference guides had been written by turn-of-the-century explorers. Personnel at the Indonesian Embassy could not confirm whether people in her area continued to hunt heads or practice bone-cleansing ceremonies, but she proceeded anyway.[22] Few women raised with the creature comforts of Western life would have surrendered to such unknowns. Their shared love of animals led the Trimates to water, but it was the charismatic Louis Leakey who convinced them to drink.

Like no other field scientist save Margaret Mead, Leakey captured the attention of the American public, with his news of exotic digs and prehistoric finds. By the time he met his lady primatologists, he was already a chubby white-haired man, easily mistaken for a retired weekend golfer rather than an Indiana Jones type who excavated sites throughout the world. It may sound grandiose to describe his life's work as the quest for the physical origins of humans, but that's truly what it was. He grew increasingly convinced that answers to his questions lay in the primates that continued to roam the earth, and he piqued the interest of American sponsors about the need to observe apes in their natural habitats.[23]

But Leakey's physical ailments prevented him from conducting the long-term study of apes he had in mind. He theorized that women should do this research, for he had long admired their abilities in the field. Women, he decided, read social cues and observed the nature around them differently from men. These abilities were not learned in school, he once told Goodall. University training served only to desensitize intuition, which was why he didn't recruit primatologists from academic faculties. "He wanted someone with a mind uncluttered and unbiased by theory," she recalled, ". . . someone with a sympathetic understanding of animals. I think his reasoning was that, if you look at human mothers, they've got to have patience to be successful. Secondly, any human female must have some kind of programming to be able to understand the wants, the needs of a small creature that can't speak. . . . And thirdly, women traditionally have been responsible for keeping peace within the family. . . . And all that means a lot of patience and ability to just watch for little nonverbal signs. So that may give one an edge on looking at very complex social behavior."[24] Forty years later, Goodall remained convinced of his logic.

Leakey was a great admirer of women, though he rarely observed boundaries with them. Mary Leakey replaced his first wife after she, too, had been recruited for the field. They had three sons, and Louis's willingness to take on child-rearing responsibilities allowed Mary to dig for long stints of uninterrupted time. In 1948 she found a nearly intact skull of the extinct ape *Proconsul*, and she unearthed numbers of prehistoric tools and animals before discovering the jaw of *Zinjanthropus*, the "Nutcracker Man," in 1959. With Mary firmly ensconced at Olduvai, Leakey turned his attention to younger women who could carry out his primate studies in the field. Mary resented his exploits and believed his "primate ladies" were nothing more than dispensers of affirmation for a man sapped of virility. She had seen it all before.

Goodall was only twenty-three when she met Leakey, Fossey was thirty-one, Galdikas twenty-five. "He is *so* sweet, so utterly adorable," Goodall wrote in 1957, though he became less so as she continually rebuffed his romantic overtures. She was able ultimately to channel Leakey's feelings into father-like affection, but Fossey received love letters from him until his death in 1972. Galdikas thought his adolescent crushes harmless, even endearing. "Louis craved female attention and warmth," she explained. He truly loved women, "but it was 'women' in the collective more than the particular."[25]

Regardless of how one might judge him, Leakey's special fondness for women brought them to the field and propelled primatology into public consciousness. The social astuteness he observed in women had little to do with any stereotyped proclivity to make small talk at cocktail parties; it was an ability to notice the smallest details while assessing the larger community or system. He made this observation at the same time that David Botstein was noting similar powers in Barbara McClintock at Cold Spring Harbor. As a roomful of people listened to papers, she quietly observed and decided that Botstein must be hard of hearing. He asked her how on earth she could have known that, to which she told him that she noted his preference for sitting in the back row of the auditorium even though he was talkative. Given the acoustics of the room, she figured he must sit in the back to hear better.[26] McClintock's perceptiveness in the lecture hall was what Leakey appreciated about women in the field. His litmus tests for choosing his researchers were brainteasers from popular magazines and a series of innocuous card games. Placing cards facedown on a coffee table, he asked Galdikas to tell him which cards were red and which were black. Although she couldn't tell him which were which, she noted that about half of them were bent and half were not. Leakey had bent the black ones, a detail noticed by his Lady Trimates but never by a man with scientific credentials. Women, he concluded, were better able to see details that did not yet appear important.[27]

Leakey believed not only that women were better observers, but also that they had more patience for such long-term studies as the ones he envisioned of primates. What required more commitment than nurturing lifelong relationships and rearing children into adulthood—feats that he assumed women accomplished more adeptly than men? He also expected that women would appear less threatening than men to the male-dominated ape communities they observed. His experiences had shown that men in the field would try to do what most had since the beginning of modern times: conquer nature and move on. Henry Nissen had observed chimpanzees in French Guinea only for two and a half months in the 1930s, and his was the longest study in the wild before Goodall's. George Schaller lasted a full year with the gorillas of Zaire, but then he moved on to lions and pandas. Fossey surpassed his logged hours within her first year in Africa, and Galdikas easily eclipsed R. K.

Davenport's eleven months with orangutans in Malaysia. By 1975 she had surpassed John MacKinnon, who had logged more field hours than anyone in the world. When she wrote up her dissertation she had more than sixty-eight hundred hours of field observation to draw from.[28]

In the lab, scientists felt pressure to produce data quickly, to be in the vanguard of theoretical and technological advancement. In the field, however, slow and steady finished the race; persistence and dedication were key qualities. Leakey knew from the start that his Trimates were committed people, and he was careful about helping them prepare for the fieldwork. When he discussed the logistics of Fossey's study with her in 1966, for example, he warned her to remove her appendix ahead of time, since no one would be able to save her if it burst in the field. He was pleased to discover that she arranged for the surgery, undaunted by his warning. Similarly, in the interest of avoiding all medical emergencies, Galdikas agreed to have her tonsils removed if such action would get her to Borneo.[29] She endured the initial years of physical discomfort and frustrating results, while her husband, Rod Brindamour, left, as Leakey predicted all men would. Although Brindamour had endured in Borneo for seven years, he couldn't help feeling like a "displaced housewife," with no career or paycheck of his own. He had been stripped of his virility in the Western sense. "Our divorce was as much a reflection of our culture and of the different ways Western men and women view the world," Galdikas later reflected:

> The archetypal Western male, Rod went to Borneo in search of adventure. . . . He was the Marlboro Man with a mission, saving the forest and the orangutans. But when you have the same adventure day after day, the exhilaration and the feeling of triumph fade. After seven and a half years, Rod felt that the laws protecting orangutans were being enforced and the reserve boundaries were secure. The job was done and it was time to move on, time to go back to "real life." I went to Indonesia for so-called "female" reasons: I wanted to help. If I had to take risks, I did. But I wasn't interested in adventure for adventure's sake. My triumph came from feeling at one with the orangutans and the forest; I exulted in the peace and the quiet. Because I wasn't looking for thrills, I never got bored. The more I knew about orangutans, the more I would be able to learn. After seven and a half years I felt even more committed than when I arrived.[30]

Galdikas believed that her accomplishments could be attributed to maternal persistence, albeit socially conditioned, not biologically based. Leakey, however, believed that such persistence was as innate as a mother orang's when protecting her young. On the one hand, he opened doors for women to experience the exotic adventures of virile turn-of-the-century naturalists. On the other, he based his beliefs in women's natural proclivities as mothers and nurturers, and feminists who believed in "difference" came to agree with him in the 1980s. Perhaps he was enlightened in the Western sense, or simply non-Western altogether. One need only be reminded of his advice to Galdikas to understand the difference: She recalled that he approved of her practicing birth control in the field, but insisted that in his Kikuyu experience painful clitorectomy proved the surest form—it had deterred women from engaging in sexual activity for centuries.[31]

James Krasner argued, more skeptically, that Leakey's choice of women over men in the field was gimmick more than anything else: "Leakey no doubt understood that readers [of National Geographic] who would not be interested in evolutionary theory or animal behavior would be arrested by photographs of middle-class white women embracing apes."[32] Regardless of motives to innovate or titillate, it's remarkable that Leakey acted to appoint women, given the potential risks at stake. What if his Trimates reported observations that the scientific establishment considered highly improbable? What if, in their propensities as feeling women, they committed the most heinous crime of all by growing attached to the animals?

In the end, most of the dire predictions came to pass. When Jane Goodall told the primatological community in 1960 that chimps were entirely habituated to her presence, it was more than most could swallow. That same year, leading expert Vernon Reynolds, who had just returned from Uganda, had concluded it couldn't be done. Before the 1950s primatologists caught no more than brief glimpses of the animals they studied; most refused to attempt habituation altogether, shooting the animals and studying their remains back in the lab. One need only know of the experiments performed in American labs to see how unorthodox Goodall appeared. Harry Harlow, a researcher of rhesus monkeys at the University of Wisconsin, ironically designed artificial conditions in the late 1950s and early 1960s to elicit animals' most natural responses. To prove the importance of the mother-child bond, for instance, he isolated babies from the mothers to whom they instinctually clung in the wild, putting them in cages with "surrogate" parents made of terrycloth and wire or with no surrogates at all. He left monkeys alone, devoid of social interaction, in darkness in his "pit of despair" for six-week stints, to note the depression and dysfunctional behavior that resulted. Somehow his man-made experiments were supposed to provide more accurate truths than what Goodall humanely determined in the forest.[33]

Professional peers approved because the role of master controller was unquestioned in Western science. Goodall, the novice, knew no better than to proceed with a different approach, letting chimps grab food from her hand and wander into camp. "Ah-ha," her detractors exclaimed, "then her data must be tainted." Touching animals in the field was tantamount to spitting into urine samples in a biomedical lab. Goodall thought just the opposite. She blended in, wearing the drab colors of the forest and making no effort to interact with animals who didn't approach her first. She viewed her dealings with the chimps not as manipulative, but rather as being on their terms. A sympathetic colleague called Goodall's "a humble science": "She asks the animals to tell her about themselves." Others were less generous, at which point Leakey defended his protégée as a lion defends her cubs—or, as Goodall observed, as a mother chimp defended her babes.[34]

Fossey's habituation of mountain gorillas took much longer, and the awaited point of contact was too monumental for her to remain detached in the way field texts advised. "One of my first rules to visitors was 'Never touch the gorillas,'" she recalled. "This rule was occasionally broken once I learned how much gorillas loved to be tickled." Before long she belched with them, groomed their fur, and cuddled their infants like her own. Torn by heart and head, she rationalized her methods as a balance between "open" and "obscured" contacts: "Obscured contacts were especially valuable in revealing behavior that otherwise would have been inhibited by my presence. The drawback to this method was that it contributed nothing toward the habituation process. Open contacts, however, slowly helped me win the animals' acceptance. This was especially true when I learned that imitation of some of their ordinary activities such as scratching and feeding or copying their contentment vocalizations tended to put the animals at ease more rapidly than if I simply looked at them through binoculars in an attempt to disguise the potentially threatening glass eyes from the shy animals."[35]

In 1960 George Schaller had been the only researcher to glimpse mountain gorillas in their habitat; twenty years later Fossey compiled the most complete set of data in existence on wild gorillas. From nose print sketches; to spectrograms of vocalizations; to studies of diet, birthing patterns, and anatomical proportions, she left few stones unturned. Ian Redmond, her most devoted assistant, took on the unpleasant task of heading up a parasitological study, classifying and drawing the organisms found in gorilla stool samples. Colleagues approved of his empirical data but criticized Fossey's subjective analysis of gorilla behavior. She was unmoved, believing that she had to proceed as an altogether different kind of scientist, one who wore the hat of behaviorist and genealogist in addition to the traditional primatologist who brought specimens back to the lab. She trusted the knowledge she gained by observing the animals in their natural habitat and believed that her persistence bestowed continuity on her observations; in eighteen years she followed four extended families of animals over three generations, charting births, deaths, relocations, bouts of disease, and feuds among clans. She was essentially a Margaret Mead of the apes.[36]

Schaller, too, observed primate behavior in the wild, but his impulse as a Western scientist was to limit contact before immersion tainted his objectivity. As a policy, he took notes at least 150 feet away from the animals. When Fossey, however, let the animals crawl all over her, peers feared it was for her own emotional needs rather than the good of objective science. Sandy Harcourt, a primatologist at Karisoke until 1974, thought that Fossey's relations with the gorillas were inappropriate, sometimes pathetic. As poachers, scientists, and sponsors threatened to remove Fossey from her mountain, she took solace in what she interpreted as the apes' unconditional acceptance. She was known to wander alone to lookout points with no equipment. Harcourt believed that even if Fossey did witness "natural" behavior, she couldn't record anything beyond anecdote since she didn't have her charts.[37]

Leakey would have ignored the criticism were Fossey's funding not in jeopardy. Although he never intended to alter their field techniques, he made arrangements that allowed Goodall and Fossey to gain appropriate scientific certification, a kind of "union card," enrolling them in graduate programs at Cambridge University, under the renowned Sir Robert Hinde. Hinde's New Ethology was based on statistical measurements, maps, and charts, not the free-flowing description of behavior that Goodall and Fossey spoke into tape players and transcribed into their logs. It's in the numbers, he told them, not the narration; his science was, in a nutshell, without context of any kind. Fossey reluctantly kept tabs of animals' age, sex, and corresponding dung size and used the computers on campus to work up her data. Invariably, however, she settled into her own methods when she got back into the field. Goodall fell into bouts of depression as each semester began and she had to leave Tanzania for England. She was the first woman admitted to a Cambridge doctoral program without a bachelor's degree, but by the time she earned her PhD she was not convinced that formal training had informed her work for the better. Galdikas had come to the same conclusion: "In the classes I had taken at UCLA in the mid- to late sixties, budding anthropologists and primatologists learned to put their emotions aside, observe, and not interfere." This advice fell on deaf ears once she got to Borneo, for it didn't account for how she would be moved and her instincts heightened by what she observed.[38]

In the end, Leakey's Trimates turned into "professionals": he saw to it that all earned doctorates and eventually academic positions. Goodall and Galdikas spent several semesters in the 1970s as visiting professors at Stanford and Simon Fraser universities, and in 1980 Fossey left Karisoke to teach at Cornell. Still, they had their academic detractors who believed that their field methods tainted the purity of their observations. Zoologists, ethologists, primatologists, and animal behaviorists accused them of a crime more heinous to professional sensibilities than poaching for sport: anthropomorphizing their apes. They thought it inappropriate that Goodall gave chimps names rather than numbers for identification, for this practice seemed to value intimacy over emotional distance. The controversy was ironic, for famed baboon expert Irven DeVore had named his animals in 1958. Moreover, Goodall demonstrated that the practice served pragmatic needs. She differentiated between families of animals through the first letters of names. Flo's offspring, for example, were Flint, Figan, and Fifi; visitors to Gombe knew, if nothing else, that Pom and Passion were related. Graduate students also traced genealogical histories through the names of orangutans near Camp Leakey. Galdikas named the first wild animal she recognized Alice, followed by her son Andy, and then Beth, Cara, Carl, Cindy . . . Martha, Merv, and so on. Fossey designated gorilla groups with numbers, but within them animals had names like Peanuts and Icarus.[39]

The practice of naming betrayed a wholly different orientation to subject than that of gentleman behaviorists who had studied primate

"types"—the chimp male, the gorilla female, the orangutan infant—
earlier in the century. The Trimates sought to understand animals as individuals with unique traits within a larger system, much as McClintock
studied her strains of corn. The psychoanalyst Nancy Chodorow has
theorized that women's viewing of the world from the perspective of
connection, men's from detachment, may be the inherent consequence
of early identity formation: girls ultimately identify with their mothers,
their primary love-objects, while men grow detached from them. Carol
Gilligan also uses psychological theory to explain women's tendency to
individuate. Women are holistic and integrative thinkers, according to
Gilligan, but they also particularize as they carry out their ethic of care;
men generalize and sort into categories, imposing laws on nature so that
it can be easily controlled. But names not only particularize; they open
doors to infused meanings. Was it an objective practice to label animals
according to physical traits, as in the case of Goodall's Throat Pouch or
David Greybeard? What of labels bearing symbolic significance? When
Passion bore a baby in 1977, it put an end to her killing of other females'
offspring, and thus Goodall named her newborn Pax. Fossey named a
gorilla in memory of her uncle Bert; one can only begin to speculate
about the emotional baggage attached to that decision.[40]

Goodall recalled that when she began her study in 1960, "it was not
permissible—at least not in ethological circles—to talk about an animal's mind. Only humans had minds." She was referring to an animal's
emotions as well as intellect, and yet years of observation led her to
conclude that chimps were nearly human emotionally, if less so intellectually. "I have often felt like an anthropologist taking notes on a tribe
of people," she explained. They had thoughts, imaginations, feelings—
all of which were understandable so long as one learned how to bridge
the gulf in communication. It was vindicating when university labs confirmed that, genetically, chimps and humans were nearly 99 percent
alike. The data made sense, given the parallels Goodall had noted in
the field: "the long period of childhood dependency, the importance
of learning, non-verbal communication patterns, tool-using and tool-
making, cooperation in hunting, sophisticated social manipulations,
aggressive territoriality, and a variety of helping behaviours, to name a
few." Galdikas also viewed orangutans as "our relatives . . . our kin." Their
only difference, she thought, was that they had never left the "Garden of
Eden" and thus had never lost their innocence.[41]

The Trimates were so moved by the tightness of the mother-child bond in primates that they modeled primate mothers in rehabilitating orphaned apes and raising their infant boys. Much the same way in which Flint clung to Flo, Grub clung to Goodall and grew accustomed to constant cradling and caressing. Photos in *National Geographic* revealed a virtual interchangeability between infant human and primate at Gombe, Karisoke, and Camp Leakey: Binti Brindamour, Galdikas's son, took baths with primate orphans; Goodall and Grub posed in ways identical to Flo and Fifi; Fossey graced a magazine cover cradling baby apes much as Galdikas did five years later. Her instinct to mother was not Brindamour's instinct to father in the nuclear sense, but she explained that it wasn't the orangutan male's either. When it became clear that the orphaned orangutan Sugito had chosen her to become his adoptive mother, Brindamour's knee-jerk reaction was to insist on differentiation, while Sugito's was to bite and urinate on his male competition. Galdikas's response was to become the perfect primate mother, patient and protective. Despite her husband's objections, she conceded her supper to the orphan and let him accompany her in the tub. "Under Rod's suspicious gaze, Sugito became my infant. . . . There were moments when I glanced down at Sugito holding on to me, and for a split second I forgot that he wasn't human and wasn't my biological child."[42]

In this moment Galdikas felt much as McClintock had when McClintock "got down in there" to study her corn: she achieved a feeling for her organism and had been forever changed by it. Critics believed that her emotional attachment made her blind to the ways in which apes actually differed from human beings. Sometimes she agreed: "I could rattle off a list of the differences. But I had lost that gut feeling of separation, which is an integral part of Western intellectual consciousness." Asymmetries did strike her once she observed her own son playing with infant orangs. Binti had come to climb trees and gesticulate like baby Tarzan, and for a time it seemed that distinctions between primate and human infants had "virtually disappeared." But then Binti's physical and emotional development diverged from his primate friends'. She eventually banned his playmates from camp, fearing that their nearly human feelings of sibling rivalry would result in a nonhuman display of physical strength against him.[43]

Goodall thought that there was a fine line between observer and observed but that her close proximity to the chimps gave her more insight in the end. "We are not, as once we believed, separated from the rest of the animal kingdom by an unbridgeable chasm," she wrote in 1990. "Nevertheless, we must not forget, not for an instant, that even if we do not differ from the apes in kind, but only in degree, that degree is still overwhelmingly large." It was large enough for her, too, to limit the chimps' interactions with Grub, who never entered the forest without a parent. Although he had lived at Gombe since four months of age, it was increasingly clear that if given the opportunity, the chimps would likely devour him. In her efforts to study apes outside captivity, Goodall caged her son as he slept. To her horror, the chimps Pom and Passion had posed a threat to their own, killing others' chimp babies and eating them. Her descriptions of the unexplainable violence took experts aback at a UNESCO conference in Paris; many wanted to believe that violence was peculiar to the human species, that it was a result of culture, not nature. Goodall left the door open to wonder if the nature of apes was in fact the nature of humans.[44]

Women in the Wild: Changing the
Culture of Western Science

Consider popular images of women's close associations with apes—real and fictitious, in the wild and in captivity—from Digit, to Koko, to King Kong. Images of the Trimates and their animals may have had the greatest impact on how Americans imagine this relationship. Donna Haraway called it "the *National Geographic* effect." Starting in the 1960s, film crews documented the movements of Leakey's protégées, who would have otherwise worked in virtual obscurity. The popularity of NGS publications and films grew out of the lay public's desire for "naked eye science"—the yearning to experience science as adventure rather than classroom fare. The family sagas between Flo and Flint in Goodall's television specials enraptured the young Biruté Galdikas long before *Dallas* or *Dynasty* captured public attention. Vicariously, women experienced science as a hands-on practice performed outside the lab or the purview of men. On the heels of the countercultural, feminist, and antiwar movements, the Jane Goodall of film represented peace and love and seemed to be Mother Nature in the flesh. She gracefully evolved from the virginal girl next door, as seen in *Miss Goodall and the Wild Chimpanzees* (1965), to the matronly advocate for primates in captivity in *My Life with the*

Chimpanzees (1990) and in films that continue to be shown on Animal Planet in the twenty-first century. In documentaries, the Gombe River Reserve appears timeless and serene and Goodall at peace with her life of limited human contact and competition. She speaks in lay terms and evinces maternal compassion for all nature's creatures. Hers appears a humane and accessible science.[45]

NGS documentaries presented all three Trimates as "one with nature," and primatological work as blissfully maternal. Like depictions of Nobel Prize scientists as loners, these images are manipulations of sorts. Those who held the camera for the Trimates were men—ones who created facades of women's closer connection to primates and the natural world they observed. Haraway suggests that NGS portrayed a convincing reunification of Western culture—the lady Trimates, with nature—the apes and their forests, at a time when it served capitalist ends to eliminate the appearance of schisms of race, class, gender, and colonialism in third-world areas. We know from the writings of the Trimates themselves, however, that such schisms shaped their daily lives and that their maternal mystique was overblown. While such movies as *Project X* and *Gorillas in the Mist* romanticized the study of apes and anointed the women who studied them, the reality was hardly the stuff of Hollywood. Fossey scared off many graduate students by putting them on "dung duty," which some might call "diaper duty" if it weren't such a stretch. Linda Fedigan, a practicing primatologist, has dismissed "the big brown eyes hypothesis," the biological explanation for the high percentage of women in her field. Although she doesn't doubt that values defined culturally as maternal made many women the patient observers who could hear the natural world "speak to them," she believes that the media has inaccurately perpetuated the idea that women study apes out of a need to mother living things.[46]

Accurate or not, such appearances are still significant for their impact. Leakey and the Trimates understood that their fate in the field depended on popular attention to them as iconic figures, though not necessarily as scientists. That they appeared in ways that were not only distorted but also nearly contradictory suggests yet again that consumers of their stories shared no consensus about what women in the field should represent. Often they appeared to be women of fortitude and compassion, but sometimes at the expense of their personae as legitimate scientists. Biographers found it impossible to cast Dian Fossey as both a competent scientist and appealing woman. Farley Mowat's *Woman in the Mist* (1987), Alex Shoumatoff's *African Madness* (1988), and Harold T. P. Hayes's *The Dark Romance of Dian Fossey* (1990), reveal, as their titles suggest, a woman confused about her appropriate "place," hence falling into proverbial darkness. Indeed she was complex, but in the hands of men the complexity always turned tragic. "Fossey emerges as a social outcast—an irrational eccentric, a street person, an animal," James Krasner laments. "While a certain amount of antisocial eccentricity is expected from male scientists. . . . one of the crueler truths about the way in which Fossey's story has played itself out in popular media is that her status as a scientist has been eclipsed by her role as heroic madwoman confronting the primitive."[47]

The Trimates were the objects of sexist stereotypes, and yet they also featured in ways that inverted assumptions about women and Western science. One might have expected that in the 1978 *National Geographic* spread on Galdikas's work, Binti Brindamour, the baby son of Western-educated parents, would have represented the idea of "culture," and Princess, the orphan of a poached ape, the perils of "nature." And yet in this spread both babies stand together, side by side, in a tub in which they are both similarly being cleansed of dirt and conventional meanings. The accompanying text reveals that the process of "cultured man" imposing order on "nature" has been reversed. Princess, who had been acculturated by Galdikas, learned sign language and even taught baby Binti his first sign word. In fact Princess had been acculturated so deeply inside the human camp that her rehabilitation required her to forget human culture entirely before she could return to the wild. So who represented "culture" or "nature"? At Camp Leakey such dualities simply didn't apply.[48]

Galdikas described her "laboratory" as "the living one that has existed for millennia"; it had no boundaries or procedural rules. There was no commute back to the domicile at the end of the day; entire ecosystems, bodies of water, and rainforests became places and subjects of research. Fossey, too, differentiated little between work and domestic space; her seven- by ten-foot tent was a bedroom, office, bath, and drying area for clothes. She took notes on the apes as well as "everything from weather to bird and plant life," "poacher activities," and anything else that provided context. Nothing in the natural environment lay outside the parameters of observation.[49]

Without the normalizing rhythms of the laboratory, women in the field worked after hours, sometimes days at a time if that is what nature allowed. Goodall commented in her notes that day felt like night and night like day. After long stints in the trees she would return to her cabin, where assistants, usually women, typed up the observations she recorded on tape. Her mother suggested she take off one night a week to do something unrelated to research. "Even on these nights our conversation was almost entirely 'chimp,'" Jane recalled. "If our work had not also been our pleasure it is doubtful whether we would have been able to keep up the pace." At her wedding to van Lawick, Goodall viewed wildlife photographs and cut a cake with a model of David Greybeard on top of it. She returned after only three days of honeymooning to monitor the development of a newborn baby chimp. Her domestic and observational space became one when she began feeding the chimps at the provisioning station next to her marital tent. The boundaries of Western science had been eroded in almost every sense.[50]

Like the first women to enter "domestic science" programs at the turn of the century, Goodall applied lessons learned "in the lab"—in this case, the forest—to her "home" life. The transfer of knowledge was not a one-way street in the field: mother chimps were teachers as well as objects of study. Once Jane had Grub, he, too, became an object to be observed, and she documented his development, as Margaret Mead had of her daughter in the 1930s and 1940s. For both scientists home and the field were not inherently oppositional spaces, and objects of study in them might amplify the commonalities between animals, "primitive" peoples, and Westerners. Connectedness characterized their brand of observation.

In the wild Goodall didn't need to hide her pregnancy or make superhuman claims. When her infant demanded attention, she scaled down her field observation: "He still wakes early," she wrote in the summer of 1967. "I mess around with him till about 10 & then he sleeps & I do chimps until he wakes about 1.30."[51] She spent the better part of 1968 taking care of her baby, observing the chimps, and revising her dissertation. This fluidity was not achievable for women working in laboratories back in the United States; their pregnancies and children underscored boundaries separating home life and work life, domesticity and science. The decision to take family leave after the birth of a child—the decision to have a child—were gambles that could cost a career.

The sexual trysts of all three Trimates also suggest that the blurring of personal and professional relationships was inevitable in the permeable space of the field. Hugo van Lawick, whom Goodall divorced in 1974, was one of the few eligible bachelors she was in contact with, since she spent every waking moment logging hours in the forest. Leakey introduced the pair, understanding better than anyone that life in the field, with its intensity and isolation, bred instant intimacy between people already connected by common interest. It was the same intimacy that led to the engagement of several field researchers at Gombe and two of Fossey's best researchers at Karisoke, much to her resentment. In the field the social protocol of the lab went out the window. Native-born trackers and credentialed scientists stood virtually on equal footing with each other and the animals. At Camp Leakey the punishing conditions of the rainforest required the unique skill set of Galdikas's second husband, Pak Bohap. He won her admiration because of his intuitive sense of the animals; his lack of formal education was irrelevant. Goodall and Fossey maintained loose pecking orders between senior scientists, junior managers, graduate students, trackers, camp and field staff, and assistants, but they often ignored seniority based on academic degrees. Jane believed that her native field assistants knew "more about following the chimpanzees through the forest, and . . . more about their behavior, than most university students."[52]

Hence the camps of the Trimates were unique meritocracies, in which intuition and intimate knowledge of subject were valued over distanced and impersonal study. Formal education, professional connections, fast-track promotions—all things men achieved more successfully than women in the lab—held no currency in the field. Individualist ambition, the mark of Anne Roe's elite scientist in the 1950s, seemed to dissolve in the trees. One need only look at the acknowledgments in the Trimates' books to understand what they valued most; they didn't thank academic mentors, but domestic workers, nannies, assistants, secretaries, and data collectors, who would have been grunts back in the lab. Gombe was a nearly classless and collaborative space, where one's hard-earned data was shared in another's dissertation research. One woman recalled with fondness the cooperative feeling of research meetings and the dinners that preceded them; no one ate until everyone had showered, dressed, and come to the table. The mood was familial, the work interdisciplinary. Zoologists, anthropologists, ethologists, ecologists, psychiatrists, and psychologists broke down disciplinary boundaries to come to an integrated understanding of primate and human.[53]

Was the Trimates' a feminist science? The question is not necessarily the same as whether or not the Trimates were feminists themselves. Dale Peterson, editor of Goodall's published letters, noted that at any given time since she had first come to Tanzania Goodall had been seen "as the little girl who thought she could; the sweet Ophelia who dreamed of animals; the feisty feminist in a man's world; the ironic traditionalist in a woman's world; the inspired nurturer; Mother Teresa of the apes; Tarzan's better half; and so on." Most of these characterizations were ones that Goodall herself had rejected. She told a reporter in 1972 that if he wanted to find evidence of women's liberation at the Gombe Reserve, he had better keep looking. Her chimps had convinced her that the traditional gender roles of the 1950s were most natural: if women had children, they belonged with them in the home. Why was she the exception, then? She wasn't, she reasoned, since her home and work were both in the forest. She portrayed herself as a sort of stay-at-home mom, who breast-fed her son on demand, since mother chimps had shown that hands-on parenting fostered self-assurance in their young.[54]

But that's not how American women generally saw her. The timing of her celebrity coincided with their growing desire to see models of female careerism and autonomy. She set off for Tanzania on the eve of President Kennedy's convening of the Commission on the Status of Women; "Jungle Jane" seemed very much the antidote to the misery Betty Friedan had exposed in suburban homes. She looked like the mom next door, but instead of driving children to school she appeared to live alone in the wild without a safety net. Dian Fossey's public image, too, took on a life of its own. A *New Yorker* staff writer referred to her as "the prototypical gutsy lady doing her thing." By virtue of living and working in the harsh Virunga Mountains, she was presumably acting in defiance of convention.[55] It was perhaps no coincidence that filmmakers chose Sigourney Weaver to play Fossey in the 1988 film adaptation of *Gorillas in the Mist*, for in *Alien* and *Working Girl* the actress had played a headstrong, independent woman as competent as any man, if not more so. It was easy to envision Weaver among the gorillas of Africa, after watching her bust proverbial balls in boardrooms and monsters' heads on other planets.

Biruté Galdikas didn't seem to mind the feminist label: "My personal decisions to get my Ph.D., to go to Southeast Asia, to spend my life studying and rescuing orangutans, and to postpone having children were all part of a wave of the future," she reflected in 1995. She had read Germaine Greer's *The Female Eunuch* before heading to the field, and it was a great disappointment when she discovered that "orangutans were decidedly sexist," as she or Greer would have defined the term. "I had come of age during the 1960s, the decade when women in North America began saying for the first time that they were no different from men," she explained. And yet the more she tried to find similarities between male and female orangutans, the more she came up short.[56]

Women primatologists have been reluctant to accept essentialist conclusions about alleged distinctions between male and female "nature," but in the past three decades they have been thoughtful about the historical and social factors that have helped to shape their field into a women-centered science. Leakey's promotion of the Trimates attracted women to the field, but older woman scientists also provided inspiration as role models. Ruth Benedict and Margaret Mead had paved the way earlier in the twentieth century for women to imagine traveling to exotic places to observe, but a tradition of fieldwork with apes in particular also persisted through the work of Frances Burton, Suzanne Ripley, Alison Richard, Barbara Harrisson, Cheryl Knott, Penny Patterson, and Jane Lancaster. Adrienne Zihlman recalled reading Mead's *Coming of Age in Samoa* and Benedict's *Patterns of Culture* in the 1950s and finding inspiration in graduate mentor Laura Nader, who taught as she nursed a young child. Younger women could envision themselves assuming roles as scientists, wives, and mothers simultaneously, since they had seen it done before. Linda Fedigan believes that women primatologists have consciously chosen a field they perceive as an "equal opportunity science." Here they have found professional camaraderie and opportunities to conduct research with feminist implications, which is often the case in newer and marginalized specialties. Fifty years ago there was not a single woman who held a primatology PhD; in 2000, women received 78 percent of the PhDs in the field.[57]

As women observed apes in their natural habitats, they amassed evidence to refute essentialist ideas validating men's dominance over women. Jane Lancaster, for example, rejected the stereotype of the passive, dependent female in 1973, and others have supported her claims

udying matrilineal networks in which estrous females, too, show
petitiveness in the realm of reproduction. "No single pattern of ag-
gressivity, dominance, troop defense, sexual dimorphism, territoriality,
competition, or any other social behaviors exist across or even within
primate species," Ruth Bleir concluded in 1978, "—except in the wishful
thinking of male investigators." The very use of the term "harem" in describing congregates of female animals she called "androcentric fantasy."
Traditionally men studied primates according to categories of "dominant males," "peripheral males," and the reproductive unit "females and
young," but women have refigured social categories and documented
evidence of matrilineal and female-bonded primate societies.[58]

The constructs of male breadwinner and female homemaker had
been outgrowths of "man the hunter" and "woman the gatherer," but
they belied women's primatological findings too. Early hominid studies decentered home and family in explanations of human evolution;
domesticity as we know it has never been an organizing principle in the
life of an ape. Sally Linton urged her colleagues in the 1970s to challenge
the notion that human beings' "first tools" presupposed the primacy
of the technology of weapons over baby slings, containers, and other
inventions that females developed as nurturers and gatherers of food.
Sarah Hrdy offered similar challenges to Irven DeVore's widely accepted
interpretations of baboons. His description of a central hierarchy of
males competing for power and access to fertile females, she concluded,
"was a more accurate portrayal of what goes on in American graduate
schools, with the big man bringing up his protégés and sleeping with
impressionable undergraduates, than of anything that goes on in baboon society."[59]

Through the 1960s, the Trimates' unique relationships to their subjects
of study continued to raise eyebrows. Robert Hinde, for example, had
encouraged his pupils to embrace a detached, empirical approach, although a decade or more later, he accepted and even championed their
observational science. The Trimates, it seems clear now, told larger-
than-life stories about their beloved animals, thus inspiring idealistic
men and women to leave the West for work with the apes. Readers of
their books cried at the deaths of Greybeard and Digit; theirs was science made popular both by their social purpose and their ethics. Critics
worried that in narrating the lives of apes the Trimates incorrectly as-

sumed that animals and humans could make sense of the world in the same ways. Galdikas, for instance, told a *New York Times* reporter that female orangs screamed only when being raped. Her characterization seemed to project human sensibility on animal behavior, but she argued that she saw orangs as individually distinctive, if not equal, in the natural order.[60]

By the 1970s scientists and laypeople were more amenable to Galdikas's perspective, with some qualifications. Critical of the masculine bravado of the atomic age and the technical proliferation that had followed, some were ready to accept kinder, gentler scientific models, which connected to nature and which valued culturally female abilities to observe and raise questions about the natural world. Alison Jolly, whose graduate training in primatology coincided with Goodall's, agrees that paradigms changed significantly between the 1960s and the 1990s. Primatology, like science in general, accounts increasingly for individual difference, but it has also become more collaborative, integrated, and environmentally conscious. Some might describe this as a turn toward feminist science, others, a turn toward science that is more humane. Why not view both brands of science as indelibly one and the same?[61]

Notes

1. Jane Goodall, *In the Shadow of Man* (Boston: Houghton Mifflin, 1971), 5.

2. Mary Palevsky, *Atomic Fragments: A Daughter's Questions* (Berkeley: University of California Press, 2000), 238.

3. Jane Goodall, *Africa in My Blood: An Autobiography in Letters: The Early Years*, ed. Dale Peterson (Boston: Houghton Mifflin, 2000), 12–25; National Geographic Video, *Jane Goodall: My Life with the Chimpanzees* (Washington, DC: National Geographic Society, 1990); Goodall, *In the Shadow of Man*, 3–4.

4. Jennifer Lindsay, *Jane Goodall: 40 Years at Gombe; A Tribute to Four Decades of Wildlife Research, Education, and Conservation* (New York: Stewart, Tabori, and Chang, 1999), 8; Jane Goodall, *Through a Window: My Thirty Years with the Chimpanzees of Gombe* (Boston: Houghton Mifflin, 1990), 5; Goodall, *In the Shadow of Man*, 40–62.

5. Jane Goodall, *Beyond Innocence: An Autobiography in Letters: The Later Years*, ed. Dale Peterson (Boston: Houghton Mifflin, 2001), 355–56; Sy Montgomery, *Walking with the Great Apes* (Boston: Houghton Mifflin, 1991), 205–6.

6. Fossey felt overshadowed by Goodall and joked that her eventual book, *Gorillas in the Mist*, should be called "In the Shadow of *In the Shadow of Man*." See Montgomery, *Walking with the Great Apes*, 149.

7. Dian Fossey, *Gorillas in the Mist* (Boston: Houghton Mifflin, 1983), 1–4; Montgomery, *Walking with the Great Apes*, 49–52; Alex Shoumatoff, *African Madness* (New York: Alfred

A. Knopf, 1988), 11–13; Farley Mowat, *Woman in the Mist: The Story of Dian Fossey and the Mountain Gorillas of Africa* (New York: Warner Books, 1987), 2–23; Harold T. P. Hayes, *The Dark Romance of Dian Fossey* (New York: Simon and Schuster, 1990), 38, 58, 114, 121–22.

8. Fossey, *Gorillas in the Mist*, xv, 2, 6; Shoumatoff, *African Madness*, 8–10, 13–14.

9. Fossey, *Gorillas in the Mist*, x, 10, 159.

10. Fossey, *Gorillas in the Mist*, 20–21, 57–58, 154–55; Shoumatoff, *African Madness*, 7, 16; Hayes, *Dark Romance*, 124–29; Mowat, *Woman in the Mist*, 58–60.

11. Shoumatoff, *African Madness*, 20–21; Fossey, *Gorillas in the Mist*, 27–28; Hayes, *Dark Romance*, 182–83, 207, 233, 295–97, 301; Mowat, *Woman in the Mist*, 59–60, 82, 90, 94, 104–5, 125, 193.

12. Hayes, *Dark Romance*, 200, 213, 216–17, 225, 229, 258; Mowat, *Woman in the Mist*, 85, 89–90, 93–94, 124, 200–1, 217, 253, 369–70; Shoumatoff, *African Madness*, 24–25, 33; Montgomery, *Walking with the Great Apes*, 53, 64–65, 133, 154, 160.

13. Fossey, *Gorillas in the Mist*, 156; Hayes, *Dark Romance*, 29; Shoumatoff, *African Madness*, 19; Mowat, *Woman in the Mist*, 155.

14. Biruté M. F. Galdikas, *Reflections of Eden: My Years with the Orangutans of Borneo* (Boston: Little, Brown, 1995), 385, 392–95.

15. Galdikas, *Reflections of Eden*, 39–43, 46–48; Evelyn Gallardo, *Among the Orangutans: The Biruté Galdikas Story* (San Francisco: Chronicle Books, 1993), 9–10; CBC Television Network, *The Third Angel* (Eugene, OR: New Dimensions Media, 1992), videocassette; Linda Spaulding, *A Dark Place in the Jungle* (Chapel Hill, NC: Algonquin Books, 1999), 65; National Geographic Video, *Search for the Great Apes* (Washington, DC: National Geographic Society, 1995), videocassette; Bettyann Kevles, *Watching the Wild Apes: The Primate Studies of Goodall, Fossey, and Galdikas* (New York: E. P. Dutton, 1976), 111–12; Montgomery, *Walking with the Great Apes*, 166–67.

16. Galdikas, *Reflections of Eden*, 83, 164.

17. Galdikas, *Reflections of Eden*, 83, 87, 90, 105–6; Gallardo, *Among the Orangutans*, 27.

18. Galdikas, *Reflections of Eden*, 300–305, 320–24, 382–84; Montgomery, *Walking with the Great Apes*, 7.

19. Spaulding, *A Dark Place in the Jungle*, 233–34.

20. Carleen Hawn, "Please Feedback the Animals," *Forbes* 170 (October 28, 2002), 168–70; Spaulding, *A Dark Place in the Jungle*, 76–77; Montgomery, *Walking with the Great Apes*, 176, 182–83.

21. Galdikas, *Reflections of Eden*, 281, 336–37.

22. Hayes, *Dark Romance*, 124–25; Galdikas, *Reflections of Eden*, 72.

23. Mary Leakey, *Disclosing the Past* (Garden City, NY: Doubleday, 1984); Montgomery, *Walking with the Great Apes*, 70–75.

24. Goodall, *In the Shadow of Man*, 6; Jocelyn Selim, "Why Chimps Still Deserve Our Respect," *Discover* 25 (May 2004), 18–19.

25. Leakey, *Disclosing the Past*, 46, 80–81, 122, 156; Goodall, *Africa in My Blood*, 109; Galdikas, *Reflections of Eden*, 65; Montgomery, *Walking with the Great Apes*, 86–87; Mowat, *Woman in the Mist*, 80–81.

26. David Botstein, "Discovery of the Bacterial Transposition Tn10," in *The Dynamic Genome: Barbara McClintock's Ideas in the Century of Genetics*, ed. Nina Federoff and David Botstein (Plainview, NY: Cold Spring Harbor Laboratory Press, 1992), 225.

27. Galdikas, *Reflections of Eden*, 49; Spaulding, *A Dark Place in the Jungle*, 65; Montgomery, *Walking with the Great Apes*, 78.

28. Fossey, *Gorillas in the Mist*, 2; Hayes, *Dark Romance*, 189; Galdikas, *Reflections of Eden*, 32, 277; Montgomery, *Walking with the Great Apes*, 80–81.

29. Fossey, *Gorillas in the Mist*, 4; Mowat, *Woman in the Mist*, 23; Gallardo, *Among the Orangutans*, 10; Montgomery, *Walking with the Great Apes*, 80.

30. Galdikas, *Reflections of Eden*, 309, 330.

31. Galdikas, *Reflections of Eden*, 300.

32. James Krasner, "'Ape Ladies' and Cultural Politics: Dian Fossey and Biruté Galdikas," in *Natural Eloquence: Women Reinscribe Science*, ed. Barbara T. Gates and Ann B. Shteir (Madison: University of Wisconsin Press, 1997), 239.

33. Deborah Blum, *The Monkey Wars* (New York: Oxford University Press, 1994), 95; H. F. Harlow, R. O. Dodsworth, and M. K. Harlow, "Total Social Isolation in Monkeys," *Proceedings of the National Academy of Sciences of the United States of America* 54 (July 1965): 90–97; Londa Schiebinger, *Has Feminism Changed Science?* (Cambridge: Harvard University Press, 1999), 6; Montgomery, *Walking with the Great Apes*, 93, 111–12.

34. Goodall, *Africa in My Blood*, 156, 190; Montgomery, *Walking with the Great Apes*, 111, 127.

35. Fossey, *Gorillas in the Mist*, 11.

36. See statistical appendices in Fossey, *Gorillas in the Mist*, 245–86.

37. Hayes, *Dark Romance*, 138–39, 198, 292–94.

38. Mowat, *Woman in the Mist*, 85–86, 88; Goodall, *Africa in My Blood*, 190–91; Montgomery, *Walking with the Great Apes*, 101–2, 105–6, 143, 147–48; Galdikas, *Reflections of Eden*, 246–47.

39. Hayes, *Dark Romance*, 140.

40. Nancy Chodorow, "Family Structure and Feminine Personality," in *Women, Culture, and Society*, ed. M. Z. Rosaldo and Louise Lamphere (Stanford: Stanford University Press, 1974), 43–66; *The Reproduction of Mothering* (Berkeley, CA: University of California Press, 1978); Carol Gilligan, *In a Different Voice: Psychological Theory and Women's Development* (Cambridge, MA: Harvard University Press, 1982); Montgomery, *Walking with the Great Apes*, 104.

41. Goodall, *Through a Window*, 14, 206; *Africa in My Blood*, 231–32; Galdikas, *Reflections of Eden*, 19.

42. Galdikas, *Reflections of Eden*, 131, 139–40, 316; Montgomery, *Walking with the Great Apes*, 35.

43. Galdikas, *Reflections of Eden*, 311–15.

44. Goodall, *Through a Window*, 207; *In the Shadow of Man*, 258–59; *Beyond Innocence*, 192.

45. Linda Marie Fedigan, "Science and the Successful Female: Why There Are So Many Women Primatologists," *American Anthropologist*, new series, 96, no. 3 (1994): 536; Donna Haraway, *Primate Visions: Gender, Race, and Nature in the World of Modern Science* (New York: Routledge, 1989), 156–60; Galdikas, *Reflections of Eden*, 30.

46. Haraway, *Primate Visions*, 150, 166, 169; Fedigan, "Science and the Successful Female," 533, 529–40; Krasner, "Ape Ladies," 237–39; Schiebinger, *Has Feminism Changed Science?* 6.

47. Krasner, "'Ape Ladies,'" 245.

48. Galdikas, "Living with Orangutans," *National Geographic*, (June 1980): 853.

49. Galdikas, *Reflections of Eden*, 20; Fossey, *Gorillas in the Mist*, 9.

50. Goodall, *In the Shadow of Man*, 132–33; *Africa in My Blood*, 271, 282.

51. Goodall, *Beyond Innocence*, 52, 73.

52. Emelie Bergman and David Riss and Geza Teleki and Ruth Davis were researchers who got engaged at Gombe. Sandy Harcourt and Kelly Stewart, daughter of actor Jimmy Stewart, got engaged at Karisoke and later married. Kevles, *Watching the Wild Apes*, 56;

Lindsay, *Jane Goodall*, 35; Goodall, *Beyond Innocence*, 2.

53. Goodall, *In the Shadow of Man*, xv–xvii; *Through a Window*, 25; *Beyond Innocence*, 145.

54. Peterson in Goodall, *Africa in My Blood*, 2; Montgomery, *Walking with the Great Apes*, 34, 40.

55. Shoumatoff, *African Madness*, 9.

56. Galdikas, *Reflections of Eden*, 330, 205.

57. Haraway, *Primate Visions*, 333; Fedigan, "Science and the Successful Female," 530–35; Schiebinger, *Has Feminism Changed Science?* 91, 127.

58. Ruth Bleier, "Bias in Biological and Human Sciences: Some Comments," *Signs* 4, no. 1 (1978): 161–62.

59. Linda Fedigan, *Primate Paradigms: Sex Roles and Social Bonds* (Montreal: Eden Press, 1982); Haraway, *Primate Visions*, 334. Hrdy quoted in Shoumatoff, *African Madness*, 19; Schiebinger, *Has Feminism Changed Science?* 127–38.

60. Mowat, *Woman in the Mist*, 269–70.

61. Montgomery, *Walking with the Great Apes*, 104.

Conclusion: Apes, Corn, and Silent Springs: A Women's Tradition of Science?

T HE RISE OF THE FEMINIST MOVEMENT AND NEW IDEAS ABOUT SCI-
entific epistemology turned the 1960s into years of flux and contradiction for women in science. The change caused discomfort for some
women, since it threatened their professional identities; the men who
had trained them had also convinced them that their teachers' perspectives of both nature and scientific practice were disinterested and ultimately authoritative. Both women and men believed that there was
no place in Western science for thoughts of ulterior or relative notions
of truth, let alone that women could uniquely achieve them. Nobelists
such as Maria Mayer and Rosalyn Yalow believed that they were scientific achievers because they operated exactly like the men around them.
To mentor other women or to support affirmative action too vigorously
might have made them appear to male colleagues to be a different kind
of scientist.

In the end Mayer never took on female students. Yalow took on female students early in her career, but eventually rebuked women of the
feminist movement. She decided that it was to her disadvantage to try
to change the scientific culture and practices she had grown to accept
and navigate skillfully. Meanwhile, other scientists, generally younger,
felt that they had much less to lose by being skeptical of the traditions
handed down to them. They began to wonder if the professional ideal
of scientific objectivity was even achievable, and some determined objectivity to be nothing more than a male epistemological stance. As for
the vast majority of women scientists, they fell between the extremes:
they sought to be scientifically disinterested and newly self-aware; to
accept the androcentric culture of science to a degree, yet to question its
traditions; to be masculine investigators, but also human beings with
empathy for living things. A life of science was a balancing act, whether
consciously perceived as such or not.

The mixed messages could have created a beleaguering crisis of identity in a woman scientist, yet for the Trimates the contradictions made it easier to conceive and practice science on new terms, to question traditional boundaries between science and domesticity, objectivity and subjectivity, nature versus human culture. They never set out to overturn gender stereotypes, yet their work challenged accepted theories about nature and the alleged naturalness of sex-typed behavior. They abjured the mechanistic mindset of Western science, believing that they could achieve understanding only by interacting with living things and observing them relating to one another in their natural habitats. *Connectedness* best characterized the Trimates' scientific philosophy, as it did Barbara McClintock's "feeling for the organism." Rachel Carson, too, was a kindred spirit, writing of the "essential unity that binds life to the earth."

Within institutional science such talk of balance with nature sounded nostalgic and quaint in the early 1960s. One reviewer of *Silent Spring* was patronizing of "Miss Carson" in 1962: "[She] maintains that the balance of nature is a major force in the survival of man, whereas the modern chemist, the modern biologist and scientist, believes that man is steadily controlling nature." Thirty years later, the environmentalist Al Gore argued that the seeming absurdity of this reviewer's worldview showed just how revolutionary Carson was; in his eyes, she was no throwback, but a visionary. As we seek green technologies and solutions for global warming and climate change in the twenty-first century, her message undoubtedly rings louder and truer than ever before. Academics once scoffed at Carson's writings for their literary (hence unscientific) qualities, but such critiques were misplaced, if for no other reason than because she brought thousands of scholarly sources to bear on her popular narratives about the environment. Today critics view her ability to give literary representation to scientific fact as one of her greatest legacies. She refused to see science as something to rarify and box off from nature, art, women, and the rest of society.[1]

Eventually few primatologists disagreed with the scientific claims the Trimates also made in their writings, although their very popularity made them suspect in the eyes of academics, much like Carson's were in the early 1960s. How could "experts" communicate so clearly to so many? Specialists scoffed, as if impermeability and truth were related. One of the greatest fictions perpetuated by the professional scientist is that his research is necessarily off limits—that one needs credentials and clearance to access it—and hence the rest of us cannot participate in it or shape it for our desired social ends. We have abetted in the deception, since we have rarely asked the scientific specialist to explain himself. We have let him proceed as if he knows better about things that affect us all.

When Anna Botsford Comstock translated the natural world into poetic and pedagogical prose in the early twentieth century, she filled important social and civic needs, equipping young people with scientific information and inspiring them to pursue lines of inquiry that they likely wouldn't have otherwise. In generations since then, women have embraced their work as scientific popularizers—Annie Cannon in the 1920s and 1930s, for example, Margaret Mead in the 1930s through the 1970s, and Lillian Gilbreth over the course of all of these decades. Science writer Diane Ackerman agrees that popularizing science is worthy work: "People want to understand the world around them, but they don't want another boring science lecture, they don't want to be talked down to, they don't want raw science offered in a way that's unrelated to their everyday lives."[2] Again, women have brought this connectedness to science, linking scientist to nonscientist and bringing extraordinary technological advances to bear on ordinary existence.

As connectors, communicators, and integrators, Carson, the Trimates, and others have been shifters of scientific paradigms, agents of revolutions, as Thomas Kuhn had once described them. They also helped to inaugurate a new type of scientific hero who bore no resemblance to the figure whom Anne Roe described in 1953. What is admirable about this scientist is not the tireless hours she sits logging data in the lab, but her ability to adapt to multiple environments—the lab, the field, the congressional floor, if necessary—and to translate her work into language that is understandable to Joe and Jane Q. Public. Unlike Roe's scientific elite, this new scientist is collaborative and multifaceted; she assumes—rather than denies—that her experience in the private realm has bearing on her science. The American public increasingly accepts this new breed of scientist and the social, political, and environmental advocacy that grows out of her sensibilities. Very few of Roe's elite men ever admitted to religious or political affiliation; they perceived themselves as above the natural or spiritual world, playing God themselves. The new type of scientist doesn't presume such power.

She is not the Ruler, but the Revolutionary—a model of grassroots action that first appealed to baby boomers coming of age with the rise of the New Left and civil rights, feminist, and antiwar movements. Al Gore was one of them, and Rachel Carson was his hero.[3]

As I have described this tradition of women in science, perhaps I have come close to sounding biologically essentialist. It is important to emphasize here that this tradition is *culturally* rather than *biologically* maternal or womanly, for women's *social* location in science and society has been the primary factor in its creation. Women are not innately closer to nature than men are. Such assumptions have followed as the historical consequence of women's being assigned to maternal status and work, as well as their being denied access to professional science, characterized throughout the twentieth century by its containment of nature in artificial settings dominated by men. As the life of Lillian Gilbreth makes clear, the categories of science and maternal domesticity have been constructed for social ends. Were they "natural," Gilbreth would not have been able to reinvent them to create new patterns of life and science.

Jane Goodall and Barbara McClintock have become especially attractive models to feminists who believe that women's biology made them better observers, but feminists of other stripes have embraced them too. Evelyn Fox Keller has described feminist science as not essentially female, but necessarily humane. Helen Longino adds that a truly feminist science is feminist in practice, not necessarily in its content or personnel. The Trimates arguably carried out both of these visions: their methods, like McClintock's, democratized scientific authority in a way that empowered the woman scientist and women and scientists at large. Some consider the Trimates' a "different" science, but their overall orientation to their subjects can also be seen as an embracing of difference *in* science, a goal of third-wave feminists in the twenty-first century.[4] Their orientation to nature, their ability to communicate their knowledge and experience of nature, and their propensity to convert their work into social and environmental advocacy—all these have been their legacies. They did not feel compelled to work through men's established routes; their outsider status freed them to see natural phenomena with their own eyes in ways that seemed novel in twentieth-century science.

Postscript: The Legacy of Madame Curie

Biographies of Lillian Gilbreth, Harriet Brooks, Lise Meitner, and Rosalind Franklin sit on my bookshelf, their legacies evaluated anew or for the first time for future generations. Even as I sit here writing in my university office, I see through the doorway a poster for a Women's History Month event honoring Rachel Carson, "a revolutionary eco-feminist for the twenty-first century." She is not being heralded as a scientist, but she is being heralded, which is more than can be said for most women who practiced science in the twentieth century. I'm heartened to see some women scientists getting belated nods of recognition, even if I sometimes question the meanings behind them. I prefer nods to women in the plural and gestures that challenge the masculinist culture of science, to celebrations of a single woman's ability to acculturate to and gain acceptance in the masculine culture.

Indeed the number of such works is growing, yet Marie Curie singularly remains the best recognized of woman scientists, living or dead. The reasons are paradoxical: she stands out as truly exceptional, and yet almost everyone sees herself in some part of Curie's legend. When she arrived in New York in 1921, Americans thought her eccentric yet acceptably maternal; by the time she left that city in 1929, public sentiments were more mixed. The exposure to radioactive materials that made legends of Curie's male peers turned her into a martyr in some circles, yet cast her in others as an irresponsible woman who endangered her unborn children for scientific exploit. Some thought her admirable in her suffering, but she *always* appeared to be suffering—necessarily and inherently suffering—since Americans viewed her as a woman trying to do things ultimately incommensurate with womanhood. As a consequence, historian Naomi Oreskes thinks that the narrative conventions of heroism came out all wrong in Curie's persona. "Her biographers generally cast her as a drudge rather than an Achilles, her physical exertions closer to those of a charwoman than a superman."[5]

Oreskes rightly identifies sexist discourse that has swirled around Curie over the years, but this discourse has also competed with counter-discourses that grow more influential over time. More than any other woman, Curie has transcended her science to become any woman's matron saint. In one context or another, bookish young girls have seen themselves in her, as have professionals, radical feminists, the religiously persecuted, the poor, the virginal, and the picked on. Remarkably, this identification has occurred despite Curie's being socially uncommitted, world renowned, sexually private, and largely silent on the question of women's rights.

When Curie died, most surmised that she had succumbed to ailments related to years of exposure to radioactive materials, a diagnosis that she privately came to accept. But consensus about what killed her has never been consensus among generations of Americans about how she reached that point. Was she too obsessive a woman—so single-minded about her science that she continued exposing herself to substances that tragically killed her? Or was she a martyr—so dedicated to good deeds that she sacrificed her life for science and humankind? Did her life's work reflect a maternal sacrifice to humanity or a mannish ambition, the likes of which we've rarely seen? The multiple ways Americans have spun and interpreted the Curie legend suggest that there has never been a uniform acceptance of feminist or antifeminist ideals, or a monolithic strain of either, for Curie to represent. Some feminists have enlisted her as a model of competence in making their case for equality in the workplace, while others have found it useful to cast her as an eccentric who thumbed her nose at science-as-usual. Curie has been a readily identifiable icon through which to make claims to the power associated with science, but also to dismantle the tenets defining it as a masculine enterprise.

That her myth has evolved and proliferated was certainly never the intention of Curie herself. She sat down to write only one brief autobiographical piece in her lifetime, the one that Missy Meloney asked her to write in the aftermath of her first American tour. Upon viewing the drafts, editors asked her to humanize her self-portrait, but she told them that she didn't feel comfortable inserting "personal incidents," especially since she wrote in tribute to her husband, not to aggrandize herself. Meloney contemplated writing a more revealing portrait, but Curie begged her to destroy all letters she had kept over the years. The pain of the Langevin scandal ran deep and caused her to prefer limited interaction with the public until she died, on Independence Day in 1934.[6]

Curie's daughters knew that their mother would have scoffed at the posthumous treatments of her. They tried to suppress them, but

Hollywood scriptwriters believed that elements of her story would appeal to Americans during the anxious years of the Depression and World War II. Shortly after her death Warner Brothers developed a Curie biopic: a story of a woman who had come from nothing to live out the American dream, albeit in the Latin Quarter of Paris. Samuel Goldwyn of MGM had a script too and assured Eve Curie that he was within his rights to create a film. Irene Curie's first impulse was to destroy her mother's papers, but Eve knew that unofficial accounts would be written anyway. She took control by writing a definitive biography and entering into an agreement with Universal Studios to make her work the basis of a screenplay. She knew little about nuclear physics and sought the counsel of scientists from the Curie Institute to write about her mother's work. Surprisingly, in the end, personal details were more elusive than professional ones. Eve's *Marie Curie* (1937) was a gesture of love as much as an attempt to right the record and reinvent her mother for future generations. She wanted her mother's story to read "like a sort of legend," and indeed that is what she achieved.[7]

Americans enlisted her mother's memory for any number of agendas in the decades to come, however, indicating that Eve didn't shut the door on claims to tell Curie's story better. Uses of her life flickered hot in the 1930s and 1940s and cold in the 1950s and 1960s and since the 1970s have kept a steady glow like the radium that made her famous. Lab notes and personal papers of hers that were off limits because of their high levels of radioactivity are now available to the public in the twenty-first century. But will they reveal better truths than those we've chosen to accept? Details of her life have been forgotten or obscured much the same way a phone message gets distorted when transferred through too many people, if transferred at all. The processes of distortion and selective memory seem to obscure the truth, but in fact they may also provide truths in themselves. As we observe how Curie has been cast and recast, we get a sense of which traits and facets of her life have struck resonant cultural chords. Although she was never more than a brief visitor to the United States in her sixty-four years of life, Americans' obsession with her is also their preoccupation with the meaning of gender and science.

In 2005 I read with great interest yet another makeover of Curie, Barbara Goldsmith's *Obsessive Genius*. I was curious about her authorial choices, for her portrait was so unlike the ones Americans first created of Curie in the 1910s and that Rosalyn Yalow idolized on the movie screen in 1943. Goldsmith's Curie, like other modern portraits, has been trimmed down to human size; she is every woman struggling to balance work and family while fending off depression, doubt, and debilitating chauvinism.[8] Her vulnerability appeals because so many of us identify with it, even if we're not scientists. It also reflects larger cultural re-definitions of scientific and female heroism taking place of late. Real women fall less short of the heroic ideal than they used to, but they still fall short, because deeply seated cultural signifiers take time to fade. Eve Curie's portrait of an altruistic, single-minded scientist and devoted wife inspired women for generations, but it also tormented women hampered by family distractions, sexual urges, and human frailties that seemed not to cross Curie's "anti-natural path."

I bring the past to bear on women's present plight in science because the Curie complex is a residue of her myth that continues to plague. Recent studies of applicants for postdoctoral fellowships indicate that women scientists' numeric scores for combined productivity and journal prestige need to be at least five times higher than men's to be considered candidates of equal merit.[9] Such findings confirm that the complex is not simply in women's heads; it continues to permeate scientific institutions and practice.

The persistent presumption that women must be more devoted, more myopic, more talented than men is a sign that modern womanhood is still defined by traditional domesticity to some degree, and modern science still defined as its antithesis. As more women enter science institutions, we cannot be content with statistics alone. We must ensure that the presumptions made, questions asked, and ends sought in these institutions reflect a more egalitarian, depolarized culture than that which has characterized most twentieth-century science. Until then, the scientist conjured in the American mind will be male, and the woman scientist an oxymoron, one who suffers conspicuousness and invisibility at once.

Notes

1. Rachel Carson, *Silent Spring* (Boston: Houghton Mifflin, 1994), xvi, xviii; Rebecca Raglon, "Rachel Carson and Her Legacy," in *Natural Eloquence: Women Reinscribe Science*, ed. Barbara T. Gates and Ann B. Shteir (Madison: University of Wisconsin Press, 1997), 198–200.

2. Barbara T. Gates and Ann B. Shteir, "Interview with Diane Ackerman, July 18, 1994," in *Natural Eloquence*, 263. Examples of current popular science writers and their works in this tradition include, Joy Adamson, *Born Free: A Lioness of Two Worlds* (New York: Pantheon, 2000); Marilynne Robinson, *Mother Country: Britain, the Welfare State and Nuclear Pollution* (New York: Farrar, Straus, and Giroux, 1989); Terry Tempest Williams, *Finding Beauty in a Broken World* (New York: Pantheon, 2008); *Red: Passion and Patience in the Desert* (New York: Vintage, 2002); *An Unspoken Hunger: Stories from the Field* (New York: Vintage, 1995).

3. Al Gore, introduction to *Silent Spring*, xviii.

4. Helen E. Longino, "Can There Be a Feminist Science?" in *Feminism and Science*, ed. Nancy Tuana (Bloomington: Indiana University Press, 1989), 45–57; Tuana, Introduction, *Feminism and Science*, viii.

5. Naomi Oreskes, "Objectivity or Heroism? On the Invisibility of Women in Science," *Osiris*, 2nd ser., 11 (1996): 110.

6. Marie Curie to Marie Meloney, July 8, 1921; September 15, 1921; August 28, 1932, Box 1, Mrs. William B. Meloney-Marie Curie Special Manuscript Collection, Columbia University Libraries, New York, NY.

7. The 1943 Hollywood film *Madame Curie* was eventually produced by MGM studios. Eve Curie to Missy Meloney, November 20, 1936; February 9, 1937; July 27, 1934; February 12, 1938, Box 1, Meloney-Curie Collection.

8. The treatments described above are characteristic of biographies written for *adults*. There is an extensive juvenile literature on Curie that continues to be compensatory in tone. Some recent adult biographies include Barbara Goldsmith, *Obsessive Genius: The Inner World of Marie Curie* (New York: W. W. Norton, 2005); Marilyn Bailey Ogilvie, *Marie Curie: A Biography* (Santa Barbara, CA: Greenwood Press, 2004); Susan Quinn, *Marie Curie: A Life* (Cambridge, MA: Da Capo Press, 1996); Rosalynd Pflaum, *Grand Obsession: Marie Curie and Her World* (New York: Doubleday, 1989).

9. Study cited in Elga Wasserman, *The Door in the Dream: Conversations with Eminent Women in Science* (Washington, DC: Joseph Henry Press, 2002), 183–84. See also Londa Schiebinger, *Has Feminism Changed Science?* (Cambridge, MA: Harvard University Press, 1999), 47–48.

Acknowledgments

VIRGINIA WOOLF LAMENTED WOMEN'S INABILITY TO CARVE OUT time and space to create, and most women understand what she meant. As I wrote this book, I took care of two babies, worked a full-teaching load, and commuted three and a half hours to work. On the nights I spent away from my kids and on train rides home, I slipped into bathrooms to plug in my breast pump and read like the dickens about women scientists. Such are the joys of juggling two children and two careers in a modern-day household. If I ever doubted that women in the past could be spread as thin, Lillian Gilbreth and other scientists in these pages set me straight.

The primatologist Jane Goodall understood this balancing act, and in her books she thanked the women who freed her from domesticity and bureaucracy long enough to do science. I, too, thank those who afforded me the luxury to think. I am grateful to Gail Curtis, Michelle Newton, and the other caregivers who watched my children as I wrote about women scientists, and I am grateful to Ana Calero for assuming many of the administrative tasks that otherwise would have reduced my time to write. When I had very little money with which to travel to archives, Dean Myrna Chase at Baruch College gave me a travel grant, and Lynda Claasen of the Mandeville Library at the University of California at San Diego helped me obtain a stipend so that I could fly across the country to learn more about Maria Goeppert Mayer. I thank them, as well as the staff of Smith College's Sophia Smith Collection, the Harvard University Archives, the Schlesinger Library, and Columbia University's Special Collections, who helped me make the most of my limited research time. Archivists are not inherently women, nor are administrative assistants and caretakers of children, but in my experience all of them were. So were the friends and family to whom I showed early chapters. I sent one rough draft on Lillian Gilbreth to my sister Jenna, who told me gently where the story started to drag; my mom, bless her heart, let me read works in progress to her over the phone. Jeanann Pannasch was ever-so patient and helpful with the later iterations of the manuscript. And my generous colleague Carol Berkin took time out from her book on Civil War women to look over some of my chapters; her model and friendship have been instrumental in my finishing this book.

In 2008 I received a teaching prize from the Mrs. Giles Whiting Foundation; rather than money, childcare, or editorial feedback, the prize provided me with time off from teaching when I most needed it to write. Dan Green took an interest in my project at its earliest stages and got my manuscript into the right hands—namely those of Florence Howe at the Feminist Press. As a historian of women, I am only too aware of what Florence has done for decades to foster better understanding of the history and literature of women all over the world. I feel privileged to benefit from her experience as an editor and feminist scholar.

I thank my husband Chris, a physicist by training and hence my appointed expert on many things technical. My kids, too, have made me acutely aware of the ways in which we cultivate and sometimes misunderstand those who someday may become the most innovative of scientific minds. And I thank my dad, who encouraged me to be curious; under different circumstances, he likely would have been a scientist.

Index

Numbers set in **bold italic** are photographs.

Firestone, Shulamith, 214–15
Fleming, Williamina Patton, 3, 88, 93, 96–100
flow chart invention, 63–64
Foot, Katherine, 93
Ford, Eleanor Clay, 44
Fossey, Dian
    anthropomorphism and, 270, 271
    biographies of, 274–75
    Biruté Galdikas and, 260
    education, 256
    photograph, 12
    research career, 257–59, 268–70, 275
Fraenkel, Elisabeth, 236–37
Franck, James, 166, 167, 169
Frankel, Mary, 143
Franklin, Christine Ladd, 39
Franklin, Rosalind
    biographies of, 232–33
    cancer and death of, 190
    education, 182
    exclusion from Nobel prize, 162, 180–81
    lack of mentors and partners, 161, 185, 186
    male colleagues and, 184, 191
    popular notions of, 191–94
    research findings, 181–82, 183, 185–90
    wartime career opportunities, 122
Frederick, Christine, 77
Freeman, Joan, 228
French, Naomi Livesay, 152
Friedan, Betty, 211, 212, 278
Frisch, Otto, 142

Gage, Matilda Joslyn, 177
Galdikas, Biruté
    anthropomorphism and, 270, 271–72
    on divorce, 266
    education, 260, 270
    on feminism, 279
    on fieldwork and research, 262–63, 275
    on Louis Leakey, 264, 266, 267
    photograph, 13
    research career, 261–63
    see also Brindamour

Gamow, George, 150, 169, 194, 238
Gaposhkin, Sergei, 108–10
gender
    cultural and social constructs of science as masculine, 239–40, 240–41
    gender inequality and, 215–16
    nature/nurture theory of gender differences, 18, 212
    terminology connotations, 18
gender roles
    collaboration of women with men, 98
    employment of women in science and, 47–48, 88–89, 95–96
    female niches in astronomy, 113
    feminist perspective on, 204–5
    job sector discrimination and, 211
    McClintock's transcendence of, 242, 244–45, 247, 248, 288
    men challenging masculine science hegemony, 207–8
    postwar gender roles, 126–28, 204–7
    views of inherent traits and, 91, 97, 98, 189–90
    women workers at Oak Ridge and, 135–36
Genes, Girls, and Gamow (Watson), 194
genetics research, 162, 242–47
"gentleman scientist," 204–5
geophysics research, 119
Gilbert, Miriam, 152
Gilbreth, Frank
    courtship and marriage to Lillian, 57–58
    death of, 53–54
    plan for twelve children, 58
    professional life, 54, 55, 57–59, 60–62, 63–64, 67–68
    structuring of children's activities, 69–72
    support of Lillian's doctoral research, 59, 60, 62–63, 65–66
Gilbreth, Frank, Jr., 53, 55, 70–71, 72, 80
Gilbreth, Jane, 73–74
Gilbreth, Lillian
    appearance at MIT symposium, 212
    Brown University PhD, 66–68

sexism
 at science conferences, 203–4
 generational differences of women
  scientists, 226, 227–38
 labeling of women and, 17–18, 161,
  183, 194–95
 Maria Mayer on, 231
 Marie Curie and, 33–35, 43–44, 289
 MIT symposium and, 211–12
 Naomi Weisstein and, 203–4
 Rosalind Franklin and, 181–82, 184,
  191–95
 Rosalyn Yalow on, 226, 233
 women in elite science and, 177–79
 see also intellectual capacity
Shapley, Harlow, 103–4, 106, 108, 109,
 111–12
Shazkin, Elky, 153
shell model theory of the nucleus,
 157–59, 162–63, 173–74, 230
Shoumatoff, Alex, 275
SIE (Society of Industrial Engineers),
 58, 73–74
Silent Spring (Carson), 208–10, 286
single women, 163–64
Sklar, Alfred, 169
Smith, Alice Kimball, 145, 146, 207
Smith College, 159
Smith Observatory, 95
social sciences, 18
Society of Women Engineers, 82
"soft" science, 18, 121
Speck, Lyda, 153
Spöner, Hertha, 168
"Sputnik effect," 206
Stadtler, Lewis, 242
Stagg Field, 137
Standard Chemical Company, 25–26, 37
Standing by and Making Do (Wilson
 and Serber), 146–47
The Stars of High Luminosity (Payne-
 Gaposhkin), 108
stellar photometry, 90, 92–93, 96–104
stereotypes. See cultural stereotypes
Stevens, Nettie, 162, 195
Stobell, Ella, 92
Stone, Eleanor, 144

Stowe, Harriet Beecher, 209
Strassman, Fritz, 160
The Structure of Scientific Revolutions
 (Kuhn), 207
success of women in science
 assimilation with male colleagues,
  19, 104–5, 138–39, 147–48, 166,
  227–38, 247
 focus on physical appearance, 176–
  77, 225
 lack of mentors and, 185
 lack of wealth and access to power,
  177–79
 masculine molding of women,
  214–16
 single vs. married women, 163–64,
  228
suffrage, women's, 25, 40
Summers, Larry, 4–5, 216
Swope, Henrietta, 112–13
"systems thinkers," 244
Szilard, Leo, 4, 15, 131, 206

Tartakoff, Helen, 175
Taylor, Frederick Winslow, 59–60, 63,
 65
Tech Area at Los Alamos, 130, 134, 140,
 142–47, 150, 207
Teller, Edward, 123–24, 138, 157–58, 172
Teller, Mici, 143
Teller, Wendy, 147
therbligs, 60, 64, 79
Thomas, M. Carey, 40
Title IX of the Education Act
 Amendments, 212
Trinity Division team at Los Alamos,
 132–33, 145, 151, 153

universities
 acceptance of women students at,
  17–18
 American Association of University
  Women, 37, 39–41
 Curie's 1921 U.S. tour and, 38
 Göttingen University, 166–67
 hiring of women at co-ed campuses,
  19